THE GLOBAL TECHNOLOGY REVOLUTION

Bio/Nano/Materials Trends and Their Synergies with Information Technology by 2015

Philip S. Antón · Richard Silberglitt · James Schneider

Prepared for the
National Intelligence Council

RAND
National Defense Research Institute

The research described in this report was prepared for the National Intelligence Council. The research was conducted in RAND's National Defense Research Institute, a federally funded research and development center supported by the Office of the Secretary of Defense, the Joint Staff, the unified commands, and the defense agencies under Contract DASW01-95-C-0069.

Library of Congress Cataloging-in-Publication Data

Anton, Philip S.
The global technology revolution : bio/nano/materials trends and their synergies with information technology by 2015 / Philip S. Anton, Richard Silberglitt, James Schneider.
p. cm.
MR-1307
Includes bibliographical references.
ISBN 0-8330-2949-5
1. Technological innovations. 2. Technology and state. 3. Information technology. I. Silberglitt, R. S. (Richard S.) II. Schneider, James, 1972– III. Title.

T173.8 .A58 2001
338.9'27—dc21

2001016075

Cover design by Maritta Tapanainen

Published 2001 by RAND
1700 Main Street, P.O. Box 2138, Santa Monica, CA 90407-2138
1200 South Hayes Street, Arlington, VA 22202-5050
RAND URL: http://www.rand.org/
To order RAND documents or to obtain additional information, contact Distribution Services: Telephone: (310) 451-7002; Fax: (310) 451-6915; Internet: order@rand.org

PREFACE

This work was sponsored by the National Intelligence Council (NIC) to inform its publication of *Global Trends 2015* (GT2015). GT2015 is a follow-on report to its 1996 document *Global Trends 2010,* which identified key factors that appeared poised to shape the world by 2010.

The NIC believed that various technologies (including information technology, biotechnology, nanotechnology (broadly defined), and materials technology) have the potential for significant and dominant global effects by 2015. The input presented in this report consists of a quick foresight into global technology trends in biotechnology, nanotechnology, and materials technology and their implications for information technology and the world in 2015. It is intended to be helpful to a broad audience, including policymakers, intelligence community analysts, and the public at large. Supporting foresight and analysis on information technology was funded and reported separately (see Hundley, et al., 2000; Anderson et al., 2000 [212, 213]).

This project was conducted in the Acquisition and Technology Policy Center of RAND's National Defense Research Institute (NDRI). NDRI is a federally funded research and development center sponsored by the Office of the Secretary of Defense, the Joint Staff, the defense agencies, and the unified commands.

The NIC provides mid-term and long-term strategic thinking and intelligence estimates for the Director of Central Intelligence and key policymakers as they pursue shifting interests and foreign policy priorities.

CONTENTS

FIGURES

TABLES

SUMMARY

Life in 2015 will be revolutionized by the growing effect of multidisciplinary technology across all dimensions of life: social, economic, political, and personal. Biotechnology will enable us to identify, understand, manipulate, improve, and control living organisms (including ourselves). The revolution of information availability and utility will continue to profoundly affect the world in all these dimensions. Smart materials, agile manufacturing, and nanotechnology will change the way we produce devices while expanding their capabilities. These technologies may also be joined by "wild cards" in 2015 if barriers to their development are resolved in time.

The results could be astonishing. Effects may include significant improvements in human quality of life and life span, high rates of industrial turnover, lifetime worker training, continued globalization, reshuffling of wealth, cultural amalgamation or invasion with potential for increased tension and conflict, shifts in power from nation states to non-governmental organizations and individuals, mixed environmental effects, improvements in quality of life with accompanying prosperity and reduced tension, and the possibility of human eugenics and cloning.

The actual realization of these possibilities will depend on a number of factors, including local acceptance of technological change, levels of technology and infrastructure investments, market drivers and limitations, and technology breakthroughs and advancements. Since these factors vary across the globe, the implementation and effects of technology will also vary, especially in developing countries. Nevertheless, the overall revolution and trends will continue through much of the developed world.

The fast pace of technological development and breakthroughs makes foresight difficult, but the technology revolution seems globally significant and quite likely.

Interacting trends in biotechnology, materials technology, and nanotechnology as well as their facilitations with information technology are discussed in this report. Additional research and coverage specific to information technology can be found in Hundley et al., 2000, and Anderson et al., 2000 [212, 213].[1]

---

[1]Bracketed numbers indicate the position of the reference in the bibliography.

THE REVOLUTION OF LIVING THINGS

Biotechnology will begin to revolutionize life itself by 2015. Disease, malnutrition, food production, pollution, life expectancy, quality of life, crime, and security will be significantly addressed, improved, or augmented. Some advances could be viewed as accelerations of human-engineered evolution of plants, animals, and in some ways even humans with accompanying changes in the ecosystem. Research is also under way to create new, free-living organisms.

The following appear to be the most significant effects and issues:

- **Increased quantity and quality of human life.** A marked acceleration is likely by 2015 in the expansion of human life spans along with significant improvements in the quality of human life. Better disease control, custom drugs, gene therapy, age mitigation and reversal, memory drugs, prosthetics, bionic implants, animal transplants, and many other advances may continue to increase human life span and improve the quality of life. Some of these advances may even improve human performance beyond current levels (e.g., through artificial sensors). We anticipate that the developed world will lead the developing world in reaping these benefits as it has in the past.
- **Eugenics and cloning.** By 2015 we may have the capability to use genetic engineering techniques to "improve" the human species and clone humans. These will be very controversial developments—among the most controversial in the entire history of mankind. It is unclear whether wide-scale efforts will be initiated by 2015, and cloning of humans may not be technically feasible by 2015. However, we will probably see at least some narrow attempts such as gene therapy for genetic diseases and cloning by rogue experimenters. The controversy will be in full swing by 2015 (if not sooner).

Thus, the revolution of biology will not come without issue and unforeseen redirections. Significant ethical, moral, religious, privacy, and environmental debates and protests are already being raised in such areas as genetically modified foods, cloning, and genomic profiling. These issues should not halt this revolution, but they will modify its course over the next 15 years as the population comes to grips with the new powers enabled by biotechnology.

The revolution of biology relies heavily on technological trends not only in the biological sciences and technology but also in microelectromechanical systems, materials, imaging, sensor, and information technology. The fast pace of technological development and breakthroughs makes foresight difficult, but advances in genomic profiling, cloning, genetic modification, biomedical engineering, disease therapy, and drug developments are accelerating.

ISSUES IN BIOTECHNOLOGY

Despite these potentials, we anticipate continuing controversy over such issues as:

- Eugenics;
- Cloning of humans, including concerns over morality, errors, induced medical problems, gene ownership, and human breeding;
- Gene patents and the potential for either excessive ownership rights of sequences or insufficient intellectual property protections to encourage investments;
- The safety and ethics of genetically modified organisms;
- The use of stem cells (whose current principal source is human embryos) for tissue engineering;
- Concerns over animal rights brought about by transplantation from animals as well as the risk of trans-species disease;
- Privacy of genetic profiles (e.g., nationwide police databases of DNA profiles, denial of employment or insurance based on genetic predispositions);
- The danger of environmental havoc from genetically modified organisms (perhaps balanced by increased knowledge and control of modification functions compared to more traditional manipulation mechanisms);
- An increased risk of engineered biological weapons (perhaps balanced by an increased ability to engineer countermeasures and protections).

Nevertheless, biomedical advances (combined with other health improvements) will continue to increase human life span in those countries where they are applied. Such advances are likely to lengthen individual productivity but also will accentuate such issues as shifts in population age, financial support for retired people, and increased health care costs for individuals.

THE REVOLUTION OF MATERIALS, DEVICES, AND MANUFACTURING

Materials technology will produce products, components, and systems that are smaller, smarter, multi-functional, environmentally compatible, more survivable, and customizable. These products will not only contribute to the growing revolutions of information and biology but will have additional effects on manufacturing, logistics, and personal lifestyles.

Smart Materials

Several different materials with sensing and actuation capabilities will increasingly be used to combine these capabilities in response to environmental conditions. Applications that can be foreseen include:

- Clothes that respond to weather, interface with information systems, monitor vital signs, deliver medicines, and protect wounds;
- Personal identification and security systems; and
- Buildings and vehicles that automatically adjust to the weather.

Increases in materials performance for power sources, sensing, and actuation could also enable new and more sophisticated classes of robots and remotely guided vehicles, perhaps based on biological models.

Agile Manufacturing

Rapid prototyping, together with embedded sensors, has provided a means for accelerated and affordable design and development of complex components and systems. Together with flexible manufacturing methods and equipment, this could enable the transition to agile manufacturing systems that by 2015 will facilitate the development of global business enterprises with components more easily specified and manufactured across the globe.

Nanofabricated Semiconductors

Hardware advances for exponentially smaller, faster, and cheaper semiconductors that have fueled information technology will continue to 2015 as the transistor gate length shrinks to the deep, 20–35 nanometer scale. This trend will increase the availability of low-cost computing and enable the development of ubiquitous embedded sensors and computational systems in consumer products, appliances, and environments.

By 2015, nanomaterials such as semiconductor "quantum dots" could begin to revolutionize chemical labeling and enable rapid processing for drug discovery, blood assays, genotyping, and other biological applications.

Integrated Microsystems

Over the next 5–10 years, chemical, fluidic, optical, mechanical, and biological components will be integrated with computational logic in commercial chip designs. Instrumentation and measurement technologies are some of the most promising areas for near-term advancements and enabling effects. Biotechnology research and production, chemical synthesis, and sensors are all likely to be substantially improved by these advances by 2015. Even entire systems (such as satellites and automated laboratory processing equipment) with integrated microscale components will be built at a fraction of the cost of current macroscale systems, revolutionizing the sensing and processing of information in a variety of civilian and military applications. Advances might also enable the proliferation of some currently controlled processing capabilities (e.g., nuclear isotope separation).

TECHNOLOGY WILD CARDS

Although the technologies described above appear to have the most promise for significant global effects, such foresights are plagued with uncertainty. As time progresses, unforeseen technological developments or effects may well eclipse these trends. Other trends that because of technical challenges do not yet seem likely to have significant global effects by 2015 could become significant earlier if breakthroughs are made. Consideration of such "wild cards" helps to round out a vision of the future in which ranges of possible end states may occur.

Novel Nanoscale Computers

In the years following 2015, serious difficulties in traditional semiconductor manufacturing techniques will be reached. One potential long-term solution for overcoming obstacles to increased computational power is to shift the basis of computation to devices that take advantage of various *quantum* effects. Another approach known as *molecular electronics* would use chemically assembled logic switches organized in large numbers to form a computer. These concepts are attractive because of the huge number of parallel, low-power devices that could be developed, but they are not anticipated to have significant effects by 2015. Research will progress in these and other alternative computational paradigms in the next 15 years.

Molecular Manufacturing

A number of visionaries have advanced the concept of molecular manufacturing in which objects are assembled atom-by-atom (or molecule-by-molecule) from the bottom up (rather than from the top down using conventional fabrication techniques). Although molecular manufacturing holds the promise of significant global changes (e.g., major shifts in manufacturing technology with accompanying needs for worker retraining and opportunities for a new manufacturing paradigm in some product areas), only the most fundamental results for molecular manufacturing currently exist in isolation at the research stage. It is certainly reasonable to expect that a small-scale integrated capability could be developed over the next 15 years, but large-scale effects by 2015 are uncertain.

Self-Assembly

Though unlikely to happen on a wide scale by 2015, self-assembly methods (including the use of biological approaches) could ultimately provide a challenge to top-down semiconductor lithography and molecular manufacturing.

META-TRENDS AND IMPLICATIONS

Taken together, the revolution of information, biology, materials, devices, and manufacturing will create wide-ranging trends, concerns, and tensions across the globe by 2015.

- **Accelerating pace of technological change.** The accelerating pace of technological change combined with "creative destruction"[2] of industries will increase the importance of continued education and training. Distance learning and other alternative mechanisms will help, but such change will make it difficult for societies reluctant to change. Cultural adaptation, economic necessity, social demands, and resource availabilities will affect the scope and pace of technological adoption in each industry and society over the next 15 years. The pace and scope of such change could in turn have profound effects on the economy, society, and politics of most countries. The degree to which science and technology can accomplish such change and achieve its benefits will very much continue to depend on the will of those who create, promote, and implement it.
- **Increasingly multidisciplinary nature of technology.** Many of these technology trends are enabled by multidisciplinary contributions and interactions. Biotechnology will rely heavily on laboratory equipment providing lab-on-a-chip analysis as well as progress in bioinformatics. Microelectromechanical systems (MEMS) and smart and novel materials will enable small, ubiquitous sensors. Also, engineers are increasingly turning to biologists to understand how living organisms solve problems in dealing with a natural environment; such "biomimetic" endeavors combine the best solutions from nature with artificially engineered components to develop systems that are better than existing organisms.
- **Competition for technology development leadership.** Leadership and participation in development in each technical area will depend on a number of factors, including future regional economic arrangements (e.g., the European Union), international intellectual property rights and protections, the character of future multi-national corporations, and the role and amount of public- and private-sector research and development (R&D) investments. Currently, there are moves toward competition among regional (as opposed to national) economic alliances, increased support for a global intellectual property protection regime, more globalization, and a division of responsibilities for R&D funding (e.g., public-sector research funding with private-sector development funding).
- **Continued globalization.** Information technology, combined with its influence on other technologies (e.g., agile manufacturing), should continue to drive globalization.

---

[2]Creative destruction can be defined as "the continuous process by which emerging technologies push out the old" (Greenspan, 1999 [10]. The original use of the phrase came from Joseph A. Schumpeter's work *Capitalism, Socialism, and Democracy* (Harper & Brothers, New York, 1942, pp. 81–86).

- **Latent lateral penetration.** Older, established technologies will trickle into new markets and applications through 2015, often providing the means for the developing world to reap the benefits of technology (albeit after those countries that invest heavily in infrastructure and acquisition early on). Such penetration may involve innovation to make existing technology appropriate to new conditions and needs rather than the development of fundamentally new technology.

Concerns and Tensions

Concerns and tensions regarding the following issues already exist in many nations today and will grow over the next 15 years:

- **Class disparities.** As technology brings benefits and prosperity to its users, it may leave others behind and create new class disparities. Although technology will help alleviate some severe hardships (e.g., food shortages and nutritional problems in the developing world), it will create real economic disparities both between and within the developed and developing worlds. Those not willing or able to retrain and adapt to new business opportunities may fall further behind. Moreover, given the market weakness of poor populations in developing countries, economic incentives often will be insufficient to drive the acquisition of new technology artifacts or skills.
- **Reduced privacy.** Various threats to individual privacy include the construction of Internet-accessible databases, increased sensor capability, DNA testing, and genetic profiles that indicate disease predispositions. There is some ambivalence about privacy because of the potential benefits from these technologies (e.g., personalized products and services). Since legislation has often lagged behind the pace of technology, privacy may be addressed in reactive rather than proactive fashion with interleaving gaps in protection.
- **Cultural threats.** Many people feel that their culture's continued vitality and possibly even long-term existence may be threatened by new ways of living brought about by technology. As the benefits of technology are seen (especially by younger generations), it may be more difficult to prevent such changes even though some technologies can preserve certain cultural artifacts and values and cultural values can have an impact on guiding regulations and protections that affect technological development.

CONCLUSIONS

Beyond the agricultural and industrial revolutions of the past, a broad, multidisciplinary *technology revolution* is changing the world. Information technology is already revolutionizing our lives (especially in the developed world) and will continue to be aided by breakthroughs in materials and nanotechnology. Biotechnology will revolutionize living organisms. Materials and nanotechnology will enable the development of new devices with unforeseen capabilities. Not only are these technologies having impact on our lives, but they are heavily intertwined, making the technology revolution highly multidisciplinary and accelerating progress in each area.

The revolutionary effects of biotechnology may be the most startling. Collective breakthroughs should improve both the quality and length of human life. Engineering of the environment will be unprecedented in its degree of intervention and control. Other technology trend effects may be less obvious to the public but in hindsight may be quite revolutionary. Fundamental changes in what and how we manufacture will produce unprecedented customization and fundamentally new products and capabilities.

Despite the inherent uncertainty in looking at future trends, a range of technological possibilities and impacts are foreseeable and will depend on various enablers and barriers (see Table S.1).

These revolutionary effects are not proceeding without issue. Various ethical, economic, legal, environmental, safety, and other social concerns and decisions must be addressed as the world's population comes to grips with the potential effects these trends may have on their cultures and their lives. The most significant issues may be privacy, economic disparity, cultural threats (and reactions), and bioethics. In particular, issues such as eugenics, human cloning, and genetic modification invoke the strongest ethical and moral reactions. These issues are highly complex since they both drive technology directions and influence each other in secondary and higher-order ways. Citizens and decisionmakers need to inform themselves about technology, assembling and analyzing these complex interactions in order to truly understand the debates surrounding technology. Such steps will prevent naive decisions, maximize technology's benefit given personal values, and identify inflection points at which decisions can have the desired effect without being negated by an unanalyzed issue.

Technology's promise is here today and will march forward. It will have widespread effects across the globe. Yet, the technology revolution will not be uniform in its effect and will play out differently on the global stage depending on acceptance, investment, and a variety of other decisions. There will be no turning back, however, since some societies will avail themselves of the revolution, and globalization will thus change the environment in which each society lives. The world is in for significant change as these advances play out on the global stage.

Table S.1

The Range of Some Potential Interacting Areas and Effects of the Technology Revolution by 2015

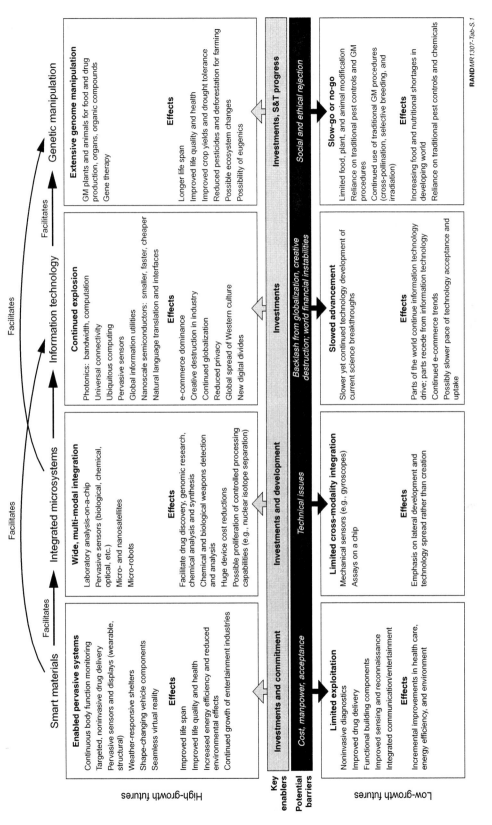

Smart materials —Facilitates→ Integrated microsystems → Information technology —Facilitates→ Genetic manipulation

(Facilitates: Smart materials → Information technology; Facilitates: Integrated microsystems → Genetic manipulation)

	Smart materials	Integrated microsystems	Information technology	Genetic manipulation
High-growth futures	**Enabled pervasive systems** Continuous body function monitoring Targeted, noninvasive drug delivery Pervasive sensors and displays (wearable, structural) Weather-responsive shelters Shape-changing vehicle components Seamless virtual reality **Effects** Improved life span Improved life quality and health Increased energy efficiency and reduced environmental effects Continued growth of entertainment industries	**Wide, multi-modal integration** Laboratory analysis-on-a-chip Pervasive sensors (biological, chemical, optical, etc.) Micro- and nanosatellites Micro-robots **Effects** Facilitate drug discovery, genomic research, chemical analysis and synthesis Chemical and biological weapons detection and analysis Huge device cost reductions Possible proliferation of controlled processing capabilities (e.g., nuclear isotope separation)	**Continued explosion** Photonics: bandwidth, computation Universal connectivity Ubiquitous computing Pervasive sensors Global information utilities Nanoscale semiconductors: smaller, faster, cheaper Natural language translation and interfaces **Effects** e-commerce dominance Creative destruction in industry Continued globalization Reduced privacy Global spread of Western culture New digital divides	**Extensive genome manipulation** GM plants and animals for food and drug production, organs, organic compounds Gene therapy **Effects** Longer life span Improved life quality and health Improved crop yields and drought tolerance Reduced pesticides and deforestation for farming Possible ecosystem changes Possibility of eugenics
Key enablers	**Investments and commitment**	**Investments and development**	**Investments**	**Investments, S&T progress**
Potential barriers	*Cost, manpower, acceptance*	*Technical issues*	*Backlash from globalization, creative destruction; world financial instabilities*	*Social and ethical rejection*
Low-growth futures	**Limited exploitation** Noninvasive diagnostics Improved drug delivery Functional building components Improved sensing and reconnaissance Integrated communication/entertainment **Effects** Incremental improvements in health care, energy efficiency, and environment	**Limited cross-modality integration** Mechanical sensors (e.g., gyroscopes) Assays on a chip **Effects** Emphasis on lateral development and technology spread rather than creation	**Slowed advancement** Slower yet continued technology development of current science breakthroughs **Effects** Parts of the world continue information technology drive; parts recede from information technology Continued e-commerce trends Possibly slower pace of technology acceptance and uptake	**Slow-go or no-go** Limited food, plant, and animal modification Reliance on traditional pest controls and GM procedures Continued use of traditional GM procedures (cross-pollination, selective breeding, and irradiation) **Effects** Increasing food and nutritional shortages in developing world Reliance on traditional pest controls and chemicals

RAND*MR1307-Tab-S.1*

ACKNOWLEDGMENTS

We would like to acknowledge the valuable insights and observations contributed by the following individuals: Robert Anderson, Jim Bonomo, Jennifer Brower, Stephan DeSpiegeleire, Bruce Don, Eugene Gritton, Richard Hundley, Eric Larson, Martin Libicki, D. J. Peterson, Steven Popper, Stephen Rattien, Calvin Shipbaugh (RAND); Claire Antón (Boeing); William Coblenz (Defense Advanced Research Projects Agency); Mark Happel (MITRE); Miguel Nicolelis (Duke University); John Pazik (Office of Naval Research); Amar Bhalla (Pennsylvania State University); Fabian Pease (Stanford University); Paul Alivisatos, Vivek Subramanian (University of California, Berkeley); Noel MacDonald (University of California, Santa Barbara); Buddy Ratner (University of Washington); Joseph Carpenter (U.S. Department of Energy); Robert Crowe (Virginia Polytechnic Institute and State University); and Lily Wu (XLinux).

Graphics production and publication were graciously facilitated by Patricia Bedrosian, Jeri Jackson, Christopher Kelly, Terri Perkins, Benson Wong, and Mary Wrazen (RAND).

Finally, we would like to thank the National Intelligence Council for its support, discussions, and encouragement throughout this project, especially Lawrence Gershwin, William Nolte, Enid Schoettle, and Brian Shaw.

ACRONYMS

AFM	Atomic-Force Microscope
BIO	Biotechnology Industry Organization
CAD	Computer-Aided Design
DoD	Department of Defense
DOE	Department of Energy
DRAMs	Dynamic Random Access Memories
FDA	Food and Drug Administration
GM	Genetically Modified
GMO	Genetically Modified Organisms
HIV	Human Immunodeficiency Virus
ITRS	International Technology Roadmap for Semiconductors
IWGN	Interagency Working Group on NanoScience
MEMS	Microelectromechanical Systems
mpg	miles per gallon
NDRI	National Defense Research Institute
NIC	National Intelligence Council
NSTC	National Science and Technology Council
PCR	Polymerase Chain Reaction
PZT	Lead Zirconate Titanate
R&D	Research and Development
S&T	Science and Technology
SPM	Scanning Probe Microscope

Chapter One

INTRODUCTION

A number of significant technology-related trends appear poised to have major global effects by 2015. These trends are being influenced by advances in biotechnology, nanotechnology,[1] materials technology, and information technology. This report presents a concise foresight[2] of these global trends and potential implications for 2015 within and among the first three technological areas as well as their intersection and cross-fertilization with information technology. This foresight activity considered potential scientific and technical advances, enabled applications, potential barriers, and global implications. These implications are varied and can include social, political, economic, environmental, or other factors. In many cases, the significance of these technologies appears to depend on the synergies afforded by their combined advances as well as on their interaction with the so-called information revolution. Unless indicated otherwise, references to possible future developments are for the 2015 timeframe.

Some have predicted that whereas the 20th century was dominated by advances in chemistry and physics, the 21st century will be dominated by advances in biotechnology (see, for example, Carey et al., 1999 [22][3]). We appear to be on the verge of understanding, reading, and controlling the genetic coding of living things, affording us revolutionary control of biological organisms and their deficiencies. Other advances in biomedical engineering, therapeutics, and drug development hold additional promises for a wide range of applications and improvements.

On another front, the U.S. President's proposed National Nanotechnology Initiative projected that "the emerging fields of nanoscience and nanoengineering are leading to unprecedented understanding and control over the fundamental building blocks of all physical things. These developments are likely to change the way almost everything—from vaccines to computers to automobile tires to objects not yet imagined—is designed and made" (National Nanotechnology Initiative, 2000 [178, 179]). This initiative reflects the optimism of many scientists who believe that technological hurdles in nanotechnology can be overcome.

---

[1]Broadly defined to include microsystems, nanosystems, and molecular systems.

[2]A foresight activity examines trends and indicators of possible future developments without predicting a single state or timeline and is thus distinct from a forecast or scenario development activity (Coates, 1985; Martin and Irvine, 1989; and Larson, 1999 [1, 2, 3]).

[3]Bracketed numbers indicate the position of the reference in the Bibliography.

In a third area, materials science and engineering is poised to provide critical inputs to both of these areas as well as creating trends of its own. For example, the cross-disciplinary fields of biomaterials (e.g., Aksay and Weiner, 1998 [131]) and nanomaterials (e.g., Lerner, 1999 [160]) are making promising developments. Moreover, interdisciplinary materials research will likely continue to yield materials with improved properties for applications that are both commonplace (such as building construction) and specialized (such as reconnaissance and surveillance, or aircraft and space systems). Materials of the 21st century[4] will likely be smarter, multi-functional, and compatible with a broad range of environments.

THE TECHNOLOGY REVOLUTION

Advances in bio/nano/materials/info technologies are combining to enable devices and systems with potential global effects on individual and public health and safety; economic, social and political systems; and business and commerce. The emerging *technology revolution*, together with the ongoing process of globalization enabled by the information technology and continued improvements in transportation (e.g., Friedman, 2000 [217]), on the one hand opens up possibilities for increased life span, economic prosperity, and quality of life, and on the other hand introduces further difficulties with privacy and ethical issues (e.g., in biomedical research). It has been argued that the accelerating pace of technological change may lead to a widening of the gap between rich and poor, developed and developing countries. However, increased global connectivity within the technology revolution may itself provide a vehicle for improved education and local technical capabilities that could enable poorer and less-developed regions of the world to contribute to and profit from technological advances via the "cottage industries" of the 21st century.

The maturity of these trends varies. Some are already producing effects and controversy in wide public forums; others hold promise for significant effects by 2015 yet are currently less mature and are mostly discussed in advanced technology forums.

APPROACH

Rather than providing a long, detailed look, this foresight activity attempted to quickly identify promising movements with potentially significant effects on the world. The study also identified "wild card" technologies that appear less promising or not likely to mature by 2015 yet would have a significant effect on the world if they were successfully developed and applied.

The determination of "global significance" in such a foresight activity depends greatly on the level at which one examines a technology or its components. Individual trends and applications may not rise to significance by themselves, but their collective contributions nevertheless might produce a significant trend. Even the Internet, for example, consists of a large number of applications, systems, and components—many of which might not hold up individually to a standard of global

[4]See, for example, Good, 1999; Arunachalam, 2000; and ASM, 2000 [124–126].

significance yet combined contribute to the overall effect. These varied contributors often come from different technical disciplines. Although multidisciplinary, such trends were grouped based on a dominant technology or a dominant concept of each trend.

Note that there is always a strong element of uncertainty when projecting technological progress and implications for the future. This effort looked for potential foreseeable implications based on progress and directions in current science and technology (S&T) and did not attempt to predict or forecast exact events and timetables. Trends were gleaned from existing outlooks, testimonies, and foresights, providing collective opinions and points of view from a broad spectrum of individuals. As many of these published trends tended to be optimistic and visionary, attempts were made to provide insights on the challenges they will face, yielding a feel not only for possible implications but also for issues that may modulate their development. The goal was to obtain a balanced perspective of current trends and directions, yielding *ranges* of possibilities rather than a single likely future to give a rich feel for the many possible paths that are being pursued. Such ranges of possible futures include both the optimistic and conservative extremes in technology foresights as well as ranges of optimistic and pessimistic implications of these trends. Some trends that hold promise but are unlikely to achieve global significance by 2015 are also mentioned.

Although the examination of trends can yield a broad understanding of current directions, it will not include unforeseen technological breakthroughs. Unforeseen complex economic, social, ethical, and political effects on technological development will also have a major effect on what actually happens in the future. For example, although many computer scientists and visionary government program managers saw the potential for the Internet[5] technology, it was practically impossible to predict whether it would become globally significant, the pace of its adoption, or its pervasive effect on social, political, and economic systems. Nevertheless, this trend study can yield a broad understanding of current issues and their potential future effects, informing policy, investment, legal, ethical, national security, and intelligence decisions today.

---

[5]Formerly called the DARPAnet developed by the Defense Advanced Research Projects Agency (DARPA).

Chapter Two

TECHNOLOGY TRENDS

GENOMICS

By 2015, biotechnology will likely continue to improve and apply its ability to profile, copy, and manipulate the genetic basis of both plants and animal organisms, opening wide opportunities and implications for understanding existing organisms and engineering organisms with new properties. Research is even under way to create new free-living organisms, initially microbes with a minimal genome (Cho et al., 1999; Hutchinson et al., 1999 [79, 80]).

Genetic Profiling and DNA Analysis

DNA analysis machines and chip-based systems will likely accelerate the proliferation of genetic analysis capabilities, improve drug search, and enable biological sensors.

The genomes of plants (ranging from important food crops such as rice and corn to production plants such as pulp trees) and animals (ranging from bacteria such as *E. coli*, through insects and mammals) will likely continue to be decoded and profiled. To the extent that genes dictate function and behavior, such extensive genetic profiling could provide an ability to better diagnose human health problems, design drugs tailored for individual problems and system reactions, better predict disease predispositions, and track disease movement and development across global populations, ethnic groups, and other genetic pools (Morton, 1999; Poste, 1999 [21, 23]). Note that a link between genes and function is generally accepted, but other factors such as the environment and phenotype play important modifying roles. Gene therapies will likely continue to be developed, although they may not mature by 2015.

Genetic profiling could also have a significant effect on security, policing, and law. DNA identification may complement existing biometric technologies (e.g., retina and fingerprint identification) for granting access to secure systems (e.g., computers, secured areas, or weapons), identifying criminals through DNA left at crime scenes, and authenticating items such as fine art. Genetic identification will likely become more commonplace tools in kidnapping, paternity, and fraud cases. Biosensors (some genetically engineered) may also aid in detecting biological warfare threats, improving food and water quality testing, continuous health monitoring, and medical laboratory analyses. Such capabilities could fundamentally change the way health services are rendered by greatly improving disease diagnosis, understanding predispositions, and improving monitoring capabilities.

Such profiling may be limited by technical difficulties in decoding some genomic segments and in understanding the implications of the genetic code. Our current technology can decode nearly all of the entire human gene sequence, but errors are still an issue, since Herculean efforts are required to decode the small amount of remaining sequences.[1] More important, although there is a strong connection between an organism's function and its genotype, we still have large gaps in understanding the intermediate steps in copying, transduction, isomer modulation, activation, immediate function, and this function's effect on larger systems in the organism. Proteomics (the study of protein function and genes) is the next big technological push after genomic decoding. Progress may likely rely on advances in bioinformatics, genetic code combination and sequencing (akin to hierarchical programming in computer languages), and other related information technologies.

Despite current optimism, a number of technical issues and hurdles could moderate genomics progress by 2015. Incomplete understanding of sequence coding, transduction, isomer modulation, activation, and resulting functions could form technological barriers to wide engineering successes. Extensive rights to own genetic codes may slow research and ultimately the benefits of the decoding. At the other extreme, the inability to secure patents from sequencing efforts may reduce commercial funding and thus slow research and resulting benefits.

In addition, investments in biotechnology have been cyclic in the past. As a result, advancements in research and development (R&D) may come in surges, especially in areas where the time to market (and thus time to return on investment) is long.

Cloning

Artificially producing genetically identical organisms through cloning will likely be significant for engineered crops, livestock, and research animals.

Cloning may become the dominant mechanism for rapidly bringing engineered traits to market, for continued maintenance of these traits, and for producing identical organisms for research and production. Research will likely continue on human cloning in unregulated parts of the world with possible success by 2015, but ethical and health concerns will likely limit wide-scale cloning of humans in regulated parts of the world. Individuals or even some states may also engage in human or animal cloning, but it is unclear what they may gain through such efforts.

---

[1]The Human Genome Project and Celera Genomics have released drafts of the human genome (IHGSC, 2001; Venter et al., 2001 [61, 64]). The drafts are undergoing additional validation, verification, and updates to weed out errors, sequence interruptions, and gaps (for details, see Pennisi, 2000, Baltimore, 2001, Aach et al., 2001, IHGSC, 2001, Galas, 2001, and Venter et al., 2001 [57, 59–61, 63, 64]). Additional technical difficulties in genomic sequencing include short, repetitive sequences that jam current DNA processing techniques as well as possible limitations of bacteria to accurately copy certain DNA fragments (Eisen, 2000; Carrington, 2000 [55, 56]).

Cloning, especially human cloning, has already generated significant controversies across the globe (Eiseman, 1999 [73]). Concerns include moral issues, the potential for errors and medical deficiencies of clones, questions of the ownership of good genes and genomes, and eugenics. Although some attempts at human cloning are possible by 2015, legal restrictions and public opinion may limit their extent. Fringe groups, however, may attempt human cloning in advance of legislative restrictions or may attempt cloning in unregulated countries. See, for example, the human cloning program announced by Clonaid (Weiss, 2000 [78]).

Although expert opinions vary regarding the current feasibility of human cloning, at least some technical hurdles for human cloning will likely need to be addressed for safe, wide-scale use. "Attempts to clone mammals from single somatic cells are plagued by high frequencies of developmental abnormalities and lethality" (Pennisi and Vogel, 2000; Matzke and Matzke, 2000 [75, 77]). Even cloned plant populations exhibit "substantial developmental and morphological irregularities" (Matzke and Matzke, 2000 [77]). Research will need to address these abnormalities or at the very least mitigate their repercussions. Some believe, however, that human cloning may be accomplished soon if the research organization accepts the high lethality rate for the embryo (Weiss, 2000 [78]) and the potential generation of developmental abnormalities.

Genetically Modified Organisms

Beyond profiling genetic codes and cloning exact copies of organisms and microorganisms, biotechnologists can also manipulate the genetic code of plants and animals and will likely continue efforts to engineer certain properties into life forms for various reasons (Long, 1998 [17]). Traditional techniques for genetic manipulation (such as cross-pollination, selective breeding, and irradiation) will likely continue to be extended by direct insertion, deletion, and modification of genes through laboratory techniques. Targets include food crops, production plants, insects, and animals.

Desirable properties could be genetically imparted to genetically engineered foods, potentially producing: improved taste; ultra-lean meats with reduced "bad" fats, salts, and chemicals; disease resistance; and artificially introduced nutrients (so-called "nutraceuticals"). Genetically modified organisms (GMOs) can potentially be engineered to improve their physical robustness, extend field and shelf life (e.g., the Flavr-Savr™ tomato[2]), tolerate herbicides, grow faster, or grow in previously unproductive environments (e.g., in high-salinity soils, with less water, or in colder climates).

Beyond systemic disease resistance, *in vivo* pesticide production has already been demonstrated (e.g., in corn) and could have a significant effect on pesticide production, application, regulation, and control with targeted release. Likewise, organisms could be engineered to produce or deliver drugs for human disease control. Cow mammary glands might be engineered to produce pharmaceuticals and therapeutic

---

[2]The Flavr-Savr trademark is held by Calgene, Inc.

organic compounds; other organisms could be engineered to produce or deliver therapeutics (e.g., the so-called "prescription banana"). If accepted by the population, such improved production and delivery mechanisms could extend the global production and availability of these therapeutics while providing easy oral delivery.

In addition to food production, plants may be engineered to improve growth, change their constitution, or artificially produce new products. Trees, for example, will likely be engineered to optimize their growth and tailor their structure for particular applications such as lumber, wood pulp for paper, fruiting, or carbon sequestering (to reduce global warming) while reducing waste byproducts. Plants might be engineered to produce bio-polymers (plastics) for engineering applications with lower pollution and without using oil reserves. Bio-fuel plants could be tailored to minimize polluting components while producing additives needed by the consuming equipment.

Genetic engineering of microorganisms has long been accepted and used. For example, *E. coli* has been used for mass production of insulin. Engineering of bacterial properties into plants and animals for disease resistance will likely occur.

Other animal manipulations could include modification of insects to impart desired behaviors, provide tagging (including GMO tagging), or prevent physical uptake properties to control pests in specific environments to improve agriculture and disease control.

Research on modifying human genes has already begun and will likely continue in a search for solutions to genetically based diseases. Although slowed by recent difficulties, gene therapy research will likely continue its search for useful mechanisms to address genetic deficiencies or for modulating physical processes such as beneficial protein production or control mechanisms for cancer. Advances in genetic profiling may improve our understanding and selection of therapy techniques and provide breakthroughs with significant health benefits.

Some cloning of humans will be possible by 2015, but legal restrictions and public opinion may limit its actual extent. Controls are also likely for human modifications (e.g., clone-based eugenic modifications) for nondisease purposes. It is possible, however, that technology will enable genetic modifications for hereditary conditions (i.e., sickle cell anemia) through *in vitro* techniques or other mechanisms.

GMOs are also having a large effect on the scientific community as an enabling technology. Not only do "knock-out" animals (animals with selected DNA sequences removed from their genome) give scientists another tool to study the effect of the removed sequence on the animal, they also enable subsequent analysis of the interaction of those functions or components with the animal's entire system. Although knock-outs are not always complete, they provide another important tool to confirm or refute hypotheses regarding complex organisms.

Broader Issues and Implications

Extant capabilities in genomics have already created opportunities yet have generated a number of issues. As more organisms are decoded and the functional implications of genes are discovered, concerns about property and privacy rights for the sequencing will likely continue.

The ability to profile an individual's DNA is already raising concerns about privacy and excessive monitoring. Examples include databases of DNA signatures for use in criminal investigations, and the potential use of genetically based health predispositions by insurance companies or employers to deny coverage or to discriminate. The latter may raise policy issues regarding acceptable and unacceptable profiling for insurance or employment. This issue is further worrisome because the exact code-to-function mechanisms that trigger many disease predispositions are not well understood.

Issues may also arise if a strong genetic basis of human physical or cognitive ability is discovered. On the positive side, understanding a person's predisposition for certain abilities (or limitations) could enable custom educational or remediation programs that will help to compensate for genetic inclinations, especially in early years when their effect can be optimized. On the negative side, groups may use such analyses in arguments to discriminate against target populations (despite, for example, the fact that ethnic distribution variances of cognitive ability are currently believed to be wider than ethnic mean differences), aggravating social and international conflicts.

Although the genetic profiles of plants have been modified for centuries using traditional techniques, questions regarding the safety of genetically modified foods have sparked international concerns in the United Kingdom and Europe, forcing a campaign by biotechnology companies to argue the safety of the technology and its applications. Some have argued that genetic engineering is actually as safe or safer than traditional combinatorial techniques such as irradiated seeds, since there often is strong supporting information concerning the function of the inserted sequences (see, for example, Somerville, 2000 [70]).

Governments have been forced into the issue, resulting in education efforts, food labeling proposals, and heated international trade discussions between the United States and Europe on the importation of GMOs and their seedlings. As genetic modification becomes more common, it may become more difficult to label and separate GMOs, resulting in a forcing function to resolve the issue of how far the technology should be applied and whether separate markets can be maintained in a global economy. This debate is starting to have global effects as populations in other countries begin to notice the impassioned debates in the United Kingdom and Europe.

Some have likened the anti-biotechnology movement to the anti-nuclear-power movement in scope and tactics, although the low cost and wide availability of basic genomic equipment and know-how will likely allow practically any country, small business, or even individual to participate in genetic engineering (Hapgood, 2000 [40]). Such wide technology availability and low entry costs could make it impossible for any movement or government to control the spread and use of genomic technology. At an extreme, successful protest pressures on big biotechnology companies together with wide technology availability could ultimately drive genomic engineering "underground" to groups outside such pressures and outside regulatory controls that help ensure safe and ethical uses. This could ironically facilitate the very problems that the anti-biotechnology movement is hoping to prevent.

Cloning and genetic modification also raise biodiversity concerns. Standardization of crops and livestock have already increased food supply vulnerabilities to diseases that can wipe out larger areas of production. Genetic modification may increase our ability to engineer responses to these threats, but the losses may still be felt in the production year unless broad-spectrum defenses are developed.

In addition to food safety, the ability to modify biological organisms holds the possibility of engineered biological weapons that circumvent current or planned countermeasures. On the other hand, genomics could aid in biological warfare defense (e.g., through improved understanding and control of biological function both in and between pathogens and target hosts as well as improved capability for engineered biosensors). Advances in genomics, therefore, could advance a race between threat engineering and countermeasures. Thus, although genetic manipulation is likely to result in medical advances, it is unclear whether we will be in a safer position in the future.

The rate at which GMO benefits are felt in poorer countries may depend on the costs of using patented organisms, marketing demands and approaches, and the rate at which crops become ubiquitous and inseparable from unmodified strains. Consider, for example, current issues related to human immunodeficiency virus (HIV) drug development and dissemination in poorer countries. Patentability has fueled research investments, but many poorer countries with dire needs cannot afford the latest drugs and must wait for handouts or patent expiration. Globalization, however, may fuel dissemination as multi-national companies invest in food production across the globe. Also, the rewards from opening previously unproductive land for production may provide the financial incentive to pay the premium for GMOs. Furthermore, widely available genomic technology could allow academics, nonprofit small businesses, and developing countries to develop GMOs to alleviate problems in poorer regions; larger biotechnology companies will focus on markets requiring capital-intensive R&D.

Finally, moral issues may play a large role in modulating the global effect of genomics trends. Some people simply believe it is improper to engineer or modify biological organisms using the new techniques. Unplanned side effects (e.g., the imposition of arthritis in current genetically modified pigs) will likely support such opposition. Others are concerned with the real danger of eugenics programs or of the engineering of dangerous biological organisms.

THERAPIES AND DRUG DEVELOPMENT

Technology

Beyond genetics, biotechnology will likely continue to improve therapies for preventing and treating disease and infection. New approaches might block a pathogen's ability to enter or travel in the body, leverage pathogen vulnerabilities, develop new countermeasure delivery mechanisms, or modulate or augment the immune response to recognizing new pathogens. These therapies may counter the current trend of increasing resistance to extant antibiotics, reshaping the war on infections.

In addition to addressing traditional viral and bacterial problems, therapies are being developed for chemical imbalances and modulation of chemical stasis. For example, antibodies are being developed that attack cocaine in the body and may be used to control addiction. Such approaches could have a significant effect on modifying the economics of the global illegal drug trade while improving conditions for users.

Drug development will likely be aided by various technology trends and enablers. Computer simulations combined with proliferating trends for molecular imaging technologies (e.g., atomic-force microscopes, mass spectroscopy, and scanning probe microscopes) may continue to improve our ability to design molecules with desired functional properties that target specific receptors, binding sites, or markers, complementing combinatorial drug search with rational drug design. Simulations of drug interactions with target biological systems could become increasing useful in understanding drug efficacy and safety. For example, Dennis Noble's complex virtual heart simulation has already contributed to U.S. Food and Drug Administration (FDA) approval of a cardiac drug by helping to understand the mechanisms and significance of an effect noticed in the clinical trial (Noble, 1998; Robbins-Roth, 1998; Buchanan, 1999 [109–111]). For some better understood systems such as the heart, this approach may become a dominant complement to clinical drug trials by 2015, whereas other more complex systems (e.g., the brain) will likely require more research on the system function and biology.

Broader Issues and Implications

R&D costs for drug development are currently extremely high and may even be unsustainable (PricewaterhouseCoopers, 1998 [19]), with averages of approximately $600 million per drug brought to market. These costs may drive the pharmaceutical industry to invest heavily in technology advances with the goal of long-term viability of the industry (PricewaterhouseCoopers, 1999 [37]). Combined with genetic profiling, drug development tailored to genotypes, chemical simulation and engineering programs, and drug testing simulations may begin to change pharmaceutical development from a broad application trial-and-error approach to custom drug development, testing, and prescription based on a deeper understanding of subpopulation response to drugs. This understanding may also rescue drugs previously rejected because of adverse reactions in small populations of clinical trials. Along with the potential for improving success rates, reducing trial costs, and opening new markets for narrowly targeted drugs, tailoring drugs to subpopulations will also have the opposite effect of reducing the size of the market for each drug. Thus, the economics of the pharmaceutical and health industries will likely change significantly if these trends come to fruition.

Note that patent protection is not uniformly enforced across the globe for the pharmaceutical industry.[3] As a result, certain regions (e.g., Asia) may continue to focus on production of non-legacy (generic) drugs, and other regions (e.g., the United States, United Kingdom, and Europe) will likely continue to pursue new drugs in addition to such low-margin pharmaceuticals.

BIOMEDICAL ENGINEERING

Multidisciplinary teaming is accelerating advances and products in biomedical engineering and technology of organic and artificial tissues, organs, and materials.

Organic Tissues and Organs

Advances in tissue and organ engineering and repair are likely to result in organic and artificial replacement parts for humans. New advances in tissue regeneration and repair continue to improve our ability to resolve health problems within our bodies.

The field of tissue engineering, which is barely a decade old, has already led to engineered commercial skin products for wound treatment.[4] Growth of cartilage for repair and replacement is at the stage of clinical testing,[5] and treatment of heart disease via growth of functional tissue by 2015 is a realistic goal.[6] These advances will depend upon improved biocompatible (or bioabsorbable) scaffold materials, development of 3D vascularized tissues and multicellular tissues, and an improved understanding of the *in vivo* growth process of cellular material on such scaffolds (Bonassar and Vacanti, 1998 [130]).

Research and applications of stem cell therapies will likely continue and expand, using these unspecialized human cells to augment or replace brain or body functions, organs (e.g., heart, kidney, liver, pancreas), and structures (Shamblott et al., 1998; Thomson et al., 1998; Couzin, 1999; Allen, 2000 [117–119, 122]). As the most unspecialized stem cells are found in early stage embryos or fetal tissue, an ethical debate is ensuing regarding the use of stem cells for research and therapy (Couzin, 1999; U.S. National Bioethics Advisory Commission, 1999; Allen, 2000 [119, 120, 122]). Alternatives such as the use of adult human stem cells or stem cell culturing may ultimately produce large-scale cell supplies with reduced ethical concerns. Current debates have limited U.S. government funding for stem cell research, but the potential has attracted substantial private funding.

---

[3] Lily Wu, personal communication.

[4] Background information and discussion of some current research can be found at http://www.pittsburgh-tissue.net and http://www.whitaker.org. Descriptions of commercial engineered skin products can be found at http://www.isotis.com http://www.advancedtissue.com, http://www.integra-ls.com, http://www.genzyme.com, and http://www.organogenesis.com.

[5] For example, see the Integra Life Sciences and Genzyme web sites above.

[6] Personal communication with Dr. Buddy Ratner, Director, University of Washington Engineered Biomaterials (UWEB) Center.

Xenotransplantations (transplantation of body parts from one species to a different species) could be improved, aided by attempts to genetically modify donor tissue and organ antibodies, complements, and regulatory proteins to reduce or eliminate rejection. Baboons or pigs, for example, may be genetically modified and cloned to produce organs for human transplant, although large-scale success may not occur by 2015.

Beyond rejection, the significance of xenotransplants is likely to be modulated by concerns that diseases such as retro viruses might jump from animals to people as a result of the transplantation techniques (Long, 1998 [17]). Ethical (e.g., animal rights) and moral concerns as well as possible patenting issues (see, for example, Walter, 1998 [208]) may also result in regulations and limitations on xenotransplants, limiting their significance.

Artificial Materials, Organs, and Bionics

In addition to organic structures, advances are likely to continue in engineering artificial tissues and organs for humans.

Multi-functional materials are being developed that provide both structure and function or that have different properties on different sides, enabling new applications and capabilities. For example, polymers with a hydrophilic shell around a hydrophobic core (biomimetic of micelles) can be used for timed release of hydrophobic drug molecules, as carriers for gene therapy or immobilized enzymes, or as artificial tissues. Sterically stabilized polymers could also be used for drug delivery.

Other materials are being developed for various biomedical applications. Fluorinated colloids, for example, are being developed that take advantage of the high electronegativity of fluorine to enhance *in vivo* oxygen transport (as a blood substitute during surgery) and for drug delivery. Hydrogels with controlled swelling behavior are being developed for drug delivery or as templates to attach growth materials for tissue engineering. Ceramics such as bioactive calcia-phosphate-silica glasses (gel-glasses), hydroxyapetite, and calcium phosphates can serve as templates for bone growth and regeneration. Bioactive polymers (e.g., polypeptides) can be applied as meshes, sponges, foams, or hydrogels to stimulate tissue growth. Coatings and surface treatments are being developed to increase biocompatibility of implanted materials (for example, to overcome the lack of endothelial cells in artificial blood vessels and reduce thrombosis). Blood substitutes may change the blood storage and retrieval systems while improving safety from blood-borne infections (Chang, 2000 [108]).

New manufacturing techniques and information technology are also enabling the production of biomedical structures with custom sizing and shape. For example, it may become commonplace to manufacture custom ceramic replacement bones for injured hands, feet, and skull parts by combining computer tomography and "rapid prototyping" (see below) to reverse engineer new bones layer by layer (Hench, 1999 [139]).

Beyond structures and organs, neural and sensor prosthetics could begin to become significant by 2015. Retinas and cochlear implants, bypasses of spinal and other nerve damage, and other artificial communications and stimulations may improve and become more commonplace and affordable, eliminating many occurrences of blindness and deafness. This could eliminate or reduce the effect of serious handicaps and change society's response from accommodation to remediation.

Biomimetics and Applied Biology

Recent techniques such as functional brain imaging and knock-out animals are revolutionizing our endeavors to understand human and animal intelligence and capabilities. These efforts should, by 2015, make significant inroads in improving our understanding of phenomena such as false memories, attention, recognition, and information processing, with implications for better understanding people and designing and interfacing artificial systems such as autonomous robots and information systems. Neuromorphic engineering (which bases its architecture and design principles on those of biological nervous systems)[7] has already produced novel control algorithms, vision chips, head-eye systems, and biomimetic autonomous robots. Although not likely to produce systems with wide intelligence or capabilities similar to those of higher organisms, this trend may produce systems by 2015 that can robustly perform useful functions such as vacuuming a house, detecting mines, or conducting autonomous search.

Surgical and Diagnostic Biotechnology

Biotechnology and materials advances are likely to continue producing revolutionary surgical procedures and systems that will significantly reduce hospital stays and cost and increase effectiveness. New surgical tools and techniques and new materials and designs for vesicle and tissue support will likely continue to reduce surgical invasiveness and offer new solutions to medical problems. Techniques such as angioplasty may continue to eliminate whole classes of surgeries; others such as laser perforations of heart tissue could promote regeneration and healing. Advances in laser surgery could refine techniques and improve human capability (e.g., LASIK[8] eye surgery to replace glasses), especially as costs are reduced and experience spreads. Hybrid imaging techniques will likely improve diagnosis, guide human and robotic surgery, and aid in basic understanding of body and brain function. Finally, collaborative information technology (e.g., "telemedicine") will likely extend specialized medical care to remote areas and aid in the global dissemination of medical quality and new advances.

---

[7]See, for example, the annual Workshop on Neuromorphic Engineering held in Telluride, Colorado (http://zig.ini.unizh.ch/telluride2000/). Mark Tilden at Los Alamos National Laboratory (funded by DARPA) has demonstrated robots that locate unexploded land mines. See the in-depth article in *Smithsonian Magazine*, February 2000, pp. 96–112. Photos of some of Tilden's robots are posted at http://www.beam-online.com/Robots/Galleria_other/tilden.html.

[8]Laser *in situ* keratomileusis.

Broader Issues and Implications

By 2015, one can envision: effective localized, targeted, and controlled drug delivery systems; long-lived implants and prosthetics; and artificial skin, bone, and perhaps heart muscle or even nerve tissue. A host of social, political, and ethical issues such as those discussed above will likely accompany these developments.

Biomedical advances (combined with other health improvements) are already increasing human life span in countries where they are applied. New advances by 2015 are likely to continue this trend, accentuating issues such as shifts in population age demographics, financial support for retired persons, and increased health care costs for individuals. Advances, however, may improve not only life expectancy but productivity and utility of these individuals, offsetting or even overcoming the resulting issues.

Many costly and specialized medical techniques are likely to initially benefit citizens who can afford better medical care (especially in developed countries, for example); wider global effects may occur later as a result of traditional trickle-down effects in medicine. Some technologies (e.g., telemedicine) may have the opposite trend where low-cost technologies may enable cost-effective consulting with specialists regardless of location. However, access to technology may greatly mediate this dispersal mechanism and may place additional demands on technology upgrades and education. Countries that remain behind in terms of technological infrastructures may miss many of these benefits.

Theological debates have also raised concerns about the definition of what constitutes a human being, since animals are being modified to produce human organs for later xenotransplantation in humans. Genetic profiling may help to inform this debate as we understand the genetic differences between humans and animals.[9]

Improved understanding of human intelligence and cognitive function could have broader legal and social effects. For example, an understanding of false memories and how they are created could have an effect on legal liabilities and courtroom testimony. Understanding innate personal capabilities and job performance requirements could help us determine who would make better fighter pilots, who has an edge in analyzing complex images,[10] and what types of improved training could improve people's capabilities to meet the special demands of their chosen careers. Ethical concerns could arise concerning discrimination against people who lack certain innate skills, requiring objective and careful measures for hiring and promotion.

Eventually, neural and sensory implants (combined with trends toward pervasive sensors in the environment and increased information availability) could radically change the way people sense, perceive, and interact with natural and artificial envi-

---

[9]For example, current estimates are that humans and chimpanzees differ genetically by only 1.5 percent (Carrington, 2000 [56]).

[10]For example, when do tetrachromats (individuals with four rather than three color detectors) have an edge and how can we identify such individuals?

ronments. Ultimately, these new capabilities could create new jobs and functions for people in these environments. Such innovations may first develop for individuals with particularly challenging and critical functions (e.g., soldiers, pilots, and controllers), but innovations may first develop in other quarters (e.g., for entertainment or business functions), given recent trends. Initial research indicates the feasibility of such implants and interactions, but it is unclear whether R&D and investments will accelerate enough to realize even such early applications by 2015. Current trends have concentrated on medical prosthetics where research prototypes are already appearing so it appears likely that globally significant systems will appear in this domain first.

THE PROCESS OF MATERIALS ENGINEERING

New materials can often be critical enabling drivers for new systems and applications with significant effects. However, it may not be obvious how enabling materials affect more observable trends and applications. A common process model from materials engineering can help to show how materials appear likely to break previous barriers in the process that ultimately results in applications with potential global benefits.

Developments in materials science and engineering result from interdisciplinary materials research. This development can be conveniently represented by the schematic description of the materials engineering process from concept to product/application (see Figure 2.1). This process view is a common approach in materials research circles and similar representations may be found in the literature (see, for example, National Research Council, 1989 [123], p. 29). Current trends in materials research that could result in global effects by 2015 are categorized below according to the process description of Figure 2.1. Figure 2.2 provides an example of the development process in the area of electroactive polymers for robotic devices and artificial muscles.

Concept/Materials Design

Biomimetics is the design of systems, materials, and their functionality to mimic nature. Current examples include layering of materials to achieve the hardness of an abalone shell or trying to understand why spider silk is stronger than steel.

Combinatorial materials design uses computing power (sometimes together with massive parallel experimentation) to screen many different materials possibilities to optimize properties for specific applications (e.g., catalysts, drugs, optical materials).

Materials Selection, Preparation, and Fabrication

Composites are combinations of metals, ceramics, polymers, and biological materials that allow multi-functional behavior. One common practice is reinforcing polymers or ceramics with ceramic fibers to increase strength while retaining light weight and

RANDMR1307-2.1

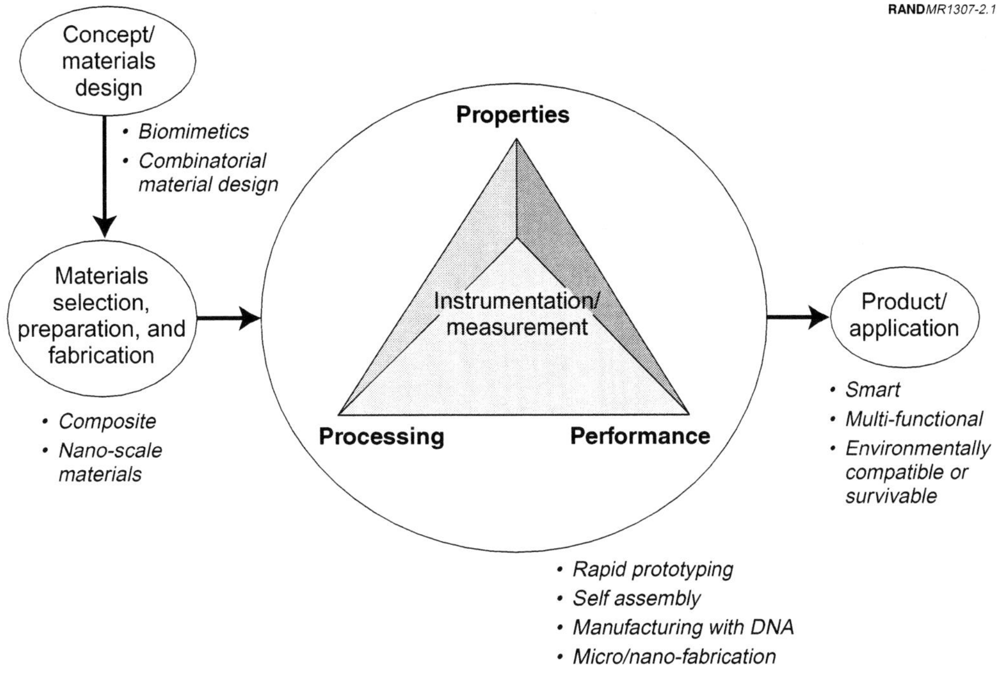

Figure 2.1—The General Materials Engineering Process

RANDMR1307-2.2

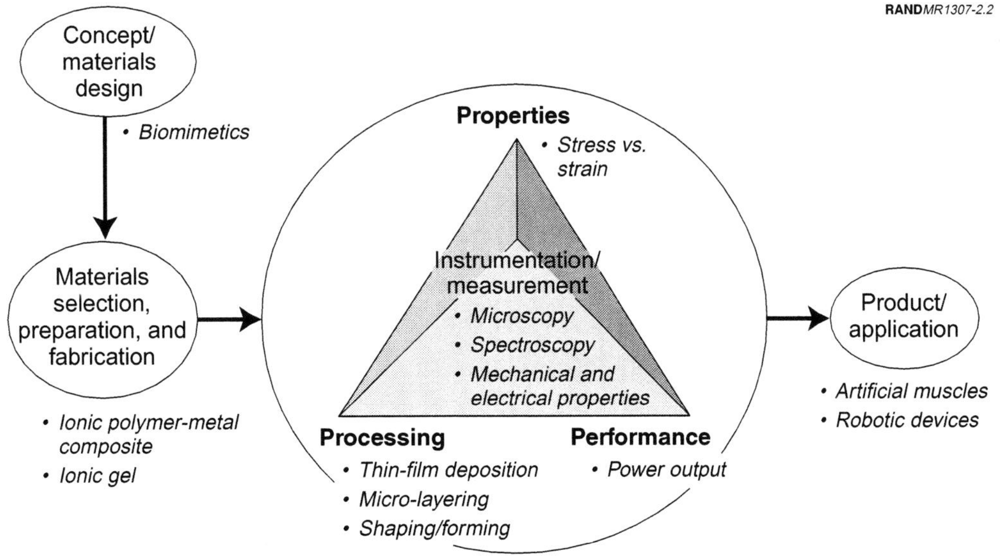

Figure 2.2—Materials Engineering Process Applied to Electroactive Polymers

avoiding the brittleness of the monolithic ceramic. Materials used in the body often combine biological and structural functions (e.g., the encapsulation of drugs).

Nanoscale materials, i.e., materials with properties that can be controlled at submicrometer ($<10^{-6}$ m) or nanometer (10^{-9} m) level, are an increasingly active area of research because properties in these size regimes are often fundamentally different from those of ordinary materials. Examples include carbon nanotubes, quantum dots, and biological molecules. These materials can be prepared either by purification methods or by tailored fabrication methods.

Processing, Properties, and Performance

These areas are inextricably linked to each other: Processing determines properties that in turn determine performance. Moreover, the sensitivity of instrumentation and measurement capability is often the enabling factor in optimizing processing, for example, as for nanotechnology and microelectromechanical systems (MEMS).

Rapid prototyping is the capability to combine computer-assisted design and manufacturing with rapid fabrication methods that allow inexpensive part production (as compared to the cost of a conventional production line). Rapid prototyping enables a company to test several different inexpensive prototypes before committing infrastructure investments to an approach. Combined with manufacturing system improvements to allow flexibility of approach and machinery, rapid prototyping can lead to an *agile manufacturing* capability. Alternatively, the company can use its virtual capability to design and then outsource product manufacturing, thus offloading capital investment and risk. This capability is synergistic with the information technology revolution in the sense that it is a further factor in globalizing manufacturing capability and enabling organizations with less capital to have a significant technological effect. For the Department of Defense (DoD), it could reduce or eliminate requirements for warehousing large amounts of spares and, for example, could enable the Air Force to "fly before they buy."

Self-assembly refers to the use in materials processing or fabrication of the tendency of some materials to organize themselves into ordered arrays (e.g., colloidal suspensions). This provides a means to achieve structured materials "from the bottom up" as opposed to using manufacturing or fabrication methods such as lithography, which is limited by the measurement and instrumentation capabilities of the day. For example, organic polymers have been tagged with dye molecules to form arrays with lattice spacing in the visible optical wavelength range and that can be changed through chemical means. This provides a material that fluoresces and changes color to indicate the presence of chemical species.

Manufacturing with DNA might represent the ultimate biomimetic manufacturing scheme. It consists of "functionalizing small inorganic building blocks with DNA and then using the molecular recognition processes associated with DNA to guide the assembly of those particles or building blocks into extended structures" (Mirkin, 2000 [106]). Using this approach, Mirkin and colleagues demonstrated a highly selective and sensitive DNA-based chemical assay method using 13 nm diameter gold particles with attached DNA sequences. This approach is compatible with the commonly used polymerase chain reaction (PCR) method of amplification of the amount of the target substance.

Micro- and nano-fabrication methods include, for example, lithography of coupled micro- or nano-scale devices on the same semiconductor or biological material. It is important to note the crucial role played in the development of these techniques by the parallel development of instrumentation and measurement devices such as the Atomic Force Microscope (AFM) and the various Scanning Probe Microscopes (SPMs).

Product/Application

The trends described above will likely work in concert to provide materials engineers with the capability to design and produce advanced materials that will be:

- *Smart*—Reactive materials combining sensors and actuators, perhaps together with computers, to enable response to environmental conditions and changes thereof. (Note, however, that limitations include the sensitivity of sensors, the performance of actuators, and the availability of power sources with required magnitude compatible with the desired size of the system.) An example might be robots that mimic insects or birds for applications such as space exploration, hazardous materials location and treatment, and unmanned aerial vehicles (UAVs).
- *Multi-functional*—MEMS and the "lab-on-a-chip" are excellent examples of systems that combine several functions. Another example is a drug delivery system using a hydrogel with hydrophilic exterior and hydrophobic interior. Consider also aircraft skins fabricated from radar-absorbing materials that incorporate avionic links and the ability to modify shape in response to airflow.
- *Environmentally compatible or survivable*—The development of composite materials and the ability to tailor materials at the atomic level will likely provide opportunities to make materials more compatible with the environments in which they will be used. Examples might include prosthetic devices that serve as templates for the growth of natural tissue and structural materials that strengthen during service (e.g., through temperature- or stress-induced phase changes).

SMART MATERIALS

Technology

Several different types of materials exhibit sensing and actuation capabilities, including ferroelectrics (exhibiting strain in response to a electric field), shape-memory alloys (exhibiting phase transition-driven shape change in response to temperature change), and magnetostrictive materials (exhibiting strain in response to a magnetic field). These effects also work in reverse, so that these materials, separately or together, can be used to combine sensing and actuation in response to environmental conditions. They are currently in widespread use in applications from ink-jet printers to magnetic disk drives to anti-coagulant devices.

An important class of smart materials is composites based upon lead zirconate titanate (PZT) and related ferroelectric materials that allow increased sensitivity, multiple frequency response, and variable frequency (Newnham, 1997 [146]). An example is the "Moonie"—a PZT transducer placed inside a half-moon-shaped cavity, which provides substantial amplification of the response. Another example is the use of composites of barium strontium titanate and non-ferroelectric materials that provide frequency-agile and field-agile responses. Applications include sensors and actuators that can change their frequency either to match a signal or to encode a signal. Ferroelectrics are already in use as nonvolatile memory elements for smart cards and as active elements in smart skis that change shape in response to stress.

Another important class of materials is smart polymers (e.g., ionic gels that deform in response to electric fields). Such electro-active polymers have already been used to make "artificial muscles" (Shahinpoor et al., 1998 [147]). Currently available materials have limited mechanical power, but this is an active research area with potential applications to robots for space exploration, hazardous duty of various types, and surveillance. Hydrogels that swell and shrink in response to changes in pH or temperature are another possibility; these hydrogels could be used to deliver encapsulated drugs in response to changes in body chemistry (e.g., insulin delivery based upon glucose concentration). Another variation on this trend for controlled release of drugs is materials with hydrophilic exterior and hydrophobic interior.

Broader Issues and Implications

A world with pervasive, networked sensors and actuators (e.g., on and part of walls, clothing, appliances, vehicles, and the environment) promises to improve, optimize, and customize the capability of systems and devices through availability of information and more direct actuation. Continuously available communication capability, ability to catalog and locate tagged personal items, and coordination of support functions have been espoused as benefits that may begin to be realized by 2015.

The continued development of small, low-profile biometric sensors, coupled with research on voice, handwriting, and fingerprint recognition, could provide effective personal security systems. These could be used for identification by police/military and also in business, personal, and leisure applications. Combined with today's information technologies, such uses could help resolve nagging security and privacy concerns while enabling other applications such as improved handgun safety (through owner identification locks) and vehicle theft control.

Other potential applications of smart materials that would be enabled by 2015 include: clothes that respond to weather, interface with information systems, monitor vital signs, deliver medicines, and automatically protect wounds; airfoils that respond to airflow; buildings that adjust to the weather; bridges and roads that sense and repair cracks; kitchens that cook with wireless instructions; virtual reality telephones and entertainment centers; and personal medical diagnostics (perhaps interfaced directly with medical care centers). The level of development and integration of these technologies into everyday life will probably depend more on consumer attitudes than on technical developments.

In addition to the surveillance and identification functions mentioned under smart materials above, developments in robotics may provide new and more sensitive capabilities for detecting and destroying explosives and contraband materials and for operating in hazardous environments. Increases in materials performance, both for power sources and for sensing and actuation, as well as integration of these functions with computing power, could enable these applications.

Such trend potentials are not without issues. Pervasive sensory information and access to collected data raise significant privacy concerns. Also, the pace of development will likely depend on investment levels and market drivers. In many cases the immediate benefits and cost savings from smart material applications will continue to drive development, but more exotic materials research may depend on public commitment to research and belief in investing in longer-term rewards.

SELF-ASSEMBLY

Technology

Examples of self-assembling materials include colloidal crystal arrays with mesoscale (50–500 nm) lattice constants that form optical diffraction gratings, and thus change color as the array swells in response to heat or chemical changes. In the case of a hydrogel with an attached side group that has molecular recognition capability, this is a chemical sensor. Self-assembling colloidal suspensions have been used to form a light-emitting diode (nanoscale), a porous metal array (by deposition followed by removal of the colloidal substrate), and a molecular computer switch.

The DNA-based self-assembly mentioned above (Mirkin, 2000 [106]) was achieved by attaching non-linking DNA strands to metal nanoparticles and adding a linking agent to form a DNA lattice. This can be turned into a biosensor or a nanolithography technique for biomolecules.

Broader Issues and Implications

Development of self-assembly methods could ultimately provide a challenge to top-down lithography approaches and molecular manufacturing approaches. As a result, it could define the next manufacturing methodology at some time beyond 2015. For example, will self-assembly methods "trump" lithography (the miracle technology of the semiconductor revolution) over the next decade or two?

RAPID PROTOTYPING

Technology

This manufacturing approach integrates computer-aided design (CAD) with rapid forming techniques to rapidly create a prototype (sometimes with embedded sensors) that can be used to visualize or test the part before making the investment in tooling required for a production run. Originally, the prototypes were made of plastic or ceramic materials and were not functional models, but now the capability exists to make a functional part, e.g., out of titanium. See, for example, the discussion of reverse-engineered bones in the section on biomedical engineering.

Broader Issues and Implications

As discussed above, agile manufacturing systems are envisioned that can connect the customer to the product throughout its life cycle and enable global business enterprises. An order would be processed using a computer-aided design, the manufacturing system would be configured in real time for the specific product (e.g., model, style, color, and options), raw materials and components would be acquired just in time, and the product would be delivered and tracked throughout its life cycle (including maintenance and recycling with identification of the customer). Components of the business enterprise could be dynamically based in the most cost-effective locations with all networked together globally. The growth of this type of business enterprise could accelerate business globalization.

BUILDINGS

Research on composite materials, waste management, and recycling has reached the stage where it is now feasible to construct buildings using materials fabricated from significant amounts of indigenous waste or recycled material content (Gupta, 2000 [127]). These approaches are finding an increasing number of cost-effective applications, especially in developing countries. Examples include the Petronas Twin Towers in Kuala Lumpur, Malaysia. These towers are the tallest buildings on earth and are made with reinforced concrete rather than steel. A roofing material used in India is made of natural fiber and agro-industrial waste. Prefabricated composite materials for home construction have also been developed in the United States, and a firm in the Netherlands is developing a potentially ubiquitous, inexpensive housing approach targeted for developing countries that uses spray-forming over an inflatable air shell.[11]

TRANSPORTATION

An important trend in transportation is the development of lightweight materials for automobiles that increase energy efficiency while reducing emissions. Here the key issue is the strength-to-weight ratio versus cost. Advanced composites with polymer, metal, or ceramic matrix and ceramic reinforcement are already in use in space systems and aircraft. These composites are too expensive for automobile applications, so aluminum alloys are being developed and introduced in cars such as the Honda Insight, the Audi A8 and AL2, and the GM EV1. Although innovation in both design

[11]For an example of the use of spray-forming over an inflatable air shell for housing, see http://www.ims.org/project/projinfo/rubacfly.htm.

and manufacturing is needed before such all-aluminum structures can become widespread, aluminum content in luxury cars and light trucks has increased in recent years. Polymer matrix, carbon-fiber (C-fiber) reinforced composites could enable high mileage cars, but C-fiber is currently several times more expensive than steel. Research sponsored by the Department of Energy (DOE) at Oak Ridge National Laboratory is working to develop cheap C-fibers which could have wider application and effect.

Spurred by California's regulations concerning ultra-low-emission vehicles, both Honda and Toyota have introduced gasoline-electric hybrid vehicles. The U.S. government and industry consortium called Partnership for a New Generation of Vehicles (PNGV) has demonstrated prototype hybrid vehicles that use both diesel/electric and diesel/fuel-cell power plants and has established 2008 as the goal for a production vehicle. These vehicles use currently available materials, but the cost reduction issues described above will be critical in bringing production costs to levels that will allow significant market penetration.

ENERGY SYSTEMS

If the ready availability of oil continues, it may be difficult for technology trends to be much of a driving force in global energy between now and 2015. Key questions have to do with continued oil imports, continued use of coal, sources of natural gas, and the fate of nuclear power. Nevertheless, technology may have significant effects in some areas.

Along with investments in solar energy, current investments in battery technology and fuel cells could enable continued trends in more portable devices and systems while extending operating times.

Developments in materials science and engineering may enable the energy systems of 2015 to be more distributed with a greater capability for energy storage, as well as energy system command, control, and communication. High-temperature superconducting cables, transformers, and storage devices could begin to increase energy transmission and distribution capabilities and power quality in this time frame.

The continued development of renewable energy could be enhanced by the combination of cheap, lightweight, recyclable materials (and perhaps the genetic engineering of biomass fuels) to provide cost-effective energy for developing countries without existing, well-developed energy infrastructures as well as for remote locations.

Significant changes in developed countries, however, may be driven more by existing social, political, and business forces, since the fuel mix of 2015 will still be strongly based on fossil fuels. Environmental concerns such as global warming and pollution might shift this direction, but it would likely require long-term economic problems (e.g., a prolonged rise in the price of oil) or distribution problems (e.g., supplies interrupted by military conflicts) to drive advances in renewable energy development.

NEW MATERIALS

Materials research may provide improvements in properties by 2015 in a number of additional areas, leading to significant effects.

SiC, GaN, and other wide band gap semiconductors are being investigated as materials for high-power electronics.

Functionally graded materials (i.e., materials whose properties change gradually from one end to the other) can form useful interlayers between mechanically, thermally, or electrically diverse components.

Anodes, cathodes, and electrolytes with higher capacity and longer lifetime are being developed for improved batteries and fuel cells.

High-temperature (ceramic) superconductors discovered in 1986 can currently operate at liquid nitrogen (rather than liquid helium) temperatures. Prototype devices such as electrical transmission cables, transformers, storage devices, motors, and fault current limiters have now been built and demonstrated. Niche application on electric utility systems should begin by 2015 (e.g., replacement of underground cables in cities and replacement of older substation transformers).

Nonlinear optical materials such as doped $LiNbO_3$ are being investigated for ultraviolet lasers (e.g., to enable finer lithography). Efforts are under way to increase damage threshold and conversion efficiency, minimize divergence, and tailor the absorption edge.

Hard materials such as nanocrystalline coatings and diamonds are being developed for applications such as computer disk drives and drill bits for oil and gas exploration, respectively.

High-temperature materials such as ductile intermetallics and ceramic matrix composites are being developed for aerospace applications and for high-efficiency energy and petrochemical conversion systems.

NANOMATERIALS

This area combines nanotechnology and many applications of nanostructured materials. One important research area is the formation of semiconductor "quantum dots" (i.e., several nanometer-size, faceted crystals) by injecting precursor materials conventionally used for chemical-vapor deposition of semiconductors into a hot liquid surfactant. This "quantum dot" is in reality a macromolecule because it is coated with a monolayer of the surfactant, preventing agglomeration. These materials photoluminesce at different frequencies (colors) depending upon their size, allowing optical multiplexing in biological labeling.[12]

---

[12]See http://www.qdots.com for a description of the applications that Quantum Dot Corporation is pursuing. Note that this approach has advantages over dyes currently in use: Quantum dots do not photobleach nearly as rapidly as dyes, enable multiplexing, and fluoresce tens of nanoseconds later than

Another important class of nanomaterials is nanotubes (the open cylindrical sisters of fullerenes).[13] Possible applications are field-emission displays (Mitsubishi research), nanoscale wires for batteries, storage of Li or H_2, and thermal management (heat pipes or insulation—the latter taking advantage of the anisotropy of thermal conductivity along and perpendicular to the tube axis). Another possibility is to use nanotubes (or fibers built from them) as reinforcement for composite materials. Presumably because of the nature of the bonding, it is predicted that nanotube-based material could be 50 to 100 times stronger than steel at one-sixth of the weight if current technical barriers can be overcome (Smalley, 1999; Service, 2000 [155, 161]).

Nanoscale structures with desirable mechanical and other properties may also be obtained through processing. Examples include strengthening of alloys with nanoscale grain structure, increased ductility of metals with multi-phase nanoscale microstructure, and increased flame retardancy of plastic nanocomposites.

NANOTECHNOLOGY

Much has been made of the trend toward producing devices with ever-decreasing scale. Many people have projected that nanometer-scale devices will continue this trend, bringing it to unprecedented levels. This includes scale reduction not only in microelectronics but also in fields such as MEMS and quantum-switch-based computing in the shorter term. These advances have the potential to change the way we engineer our environment, construct and control systems, and interact in society.

Nanofabricated Computation Devices

Nanofabricated Chips. SEMATECH—the leading industry group in the semiconductor manufacturing business—is calling for the development of nanoscale semiconductors in their latest International Technology Roadmap for Semiconductors (ITRS) (SEMATECH, 1999 [190]). The roadmap calls for a 35 nm gate length in 2015 with a total number of functions in high-volume production microprocessors of around 4.3 billion. For low-volume, high-performance processors, the number of functions may approach 20 billion. Corresponding memory chips (DRAMs) are targeted to hold around 64 gigabytes. These roadmap targets would continue the exponential trend in processing power, fueling advances in information technology. Although a number of engineering challenges exist (such as lithography, interconnects, and defect management), obstacles to achieve at least this level of performance do not seem insurmountable.

Given unforeseen shortfalls in the economic production of these chips (e.g., because of very high manufacturing costs or unacceptably large numbers of manufacturing defects), several alternatives seem possible. Defect-tolerant computer architectures

---

the auto-fluorescence (thus separating signal from noise). Thus, they may enable rapid processing for drug discovery, blood assays, genotyping, and other biological applications.

[13]For links to nanotube sites and general information, see http://www.scf.fundp.ac.be/~vmeunier/carbon_nanotube.html. Note that Professor Richard Smalley at Rice University has established a production facility (see http://cnst.rice.edu/tubes/).

such as those prototyped on a small scale by Hewlett-Packard (Heath et al., 1998 [186]) offer one alternative. These alternative methods provide some level of additional robustness to the performance goals set by the ITRS.

However, in the years following 2015, additional difficulties will likely be encountered, some of which may pose serious challenges to traditional semiconductor manufacturing techniques. In particular, limits to the degree that interconnections or "wires" between transistors may be scaled could in turn limit the effective computation speed of devices because of materials properties and compatibility, despite incremental present-day advances in these areas. Thermal dissipation in chips with extremely high device densities will also pose a serious challenge. This issue is not so much a fundamental limitation as it is an economic consideration, in that heat dissipation mechanisms and cooling technology may be required that add to total system cost, thereby adversely affecting marginal cost per computational function for these devices.

Quantum-Switch-Based Computing. One potential long-term solution for overcoming obstacles to increased computational power is computing based on devices that take advantage of various quantum effects. The core innovation in this work is the use of quantum effects, such as spin polarization of electrons, to determine the state of individual switches. This is in contrast to more traditional microelectronics, which are based on macroscopic properties of large numbers of electrons, taking advantage of materials properties of semiconductors.

Various concepts of quantum computers are attractive because of their massive parallelism in computation, but they are not anticipated to have significant effects by 2015. These concepts are qualitatively different from those employed in traditional computers and will hence require new computer architectures. The types of computations (and hence applications) that can be quickly performed using these computers are not the same as those readily addressed by today's digital computer. Several workers in the area have devised algorithms for problems that are very computationally intensive (and thus time-consuming) for existing digital computers, which could be made much faster using the physics of quantum computers. Examples of these problems include factoring large numbers (essential for cryptographic applications), searching large databases, pattern matching, and simulation of molecular and quantum phenomena.

A preliminary survey of work in this area indicates that quantum switches are unlikely to overcome major technical obstacles, such as error correction, de-coherence and signal input/output, within the next 15 years. If this were indeed the case, quantum-switch-based computing does not appear to be competitive with traditional digital electronic computers within the 2015 timeframe.

Bio-Molecular Devices and Molecular Electronics

Many of the same manufacturing and architectural challenges discussed above regarding quantum computing also hold true for molecular electronics. Molecular electronic devices could operate as logic switches through chemical means, using synthesized organic compounds. These devices can be assembled chemically in large numbers and organized to form a computer. The main advantage of this approach is significantly lower power consumption by individual devices. Several approaches for such devices have been devised, and experiments have shown evidence of switching behavior for individual devices. Several research groups have proposed interconnection between devices using carbon nanotubes, which provide high conductivity using single molecular strands of carbon. Progress has been made toward raising the operating temperature of these switches to nearly room temperature, making the switching process reversible, and increasing the overall amount of current that can be switched using these devices.

Several major outstanding issues remain with respect to molecular electronics. One issue is that molecular memories must be able to maintain their state, just as in a digital electronic computer. Also, given that the manufacturing and assembly process for these devices will lead to device defects, a defect-tolerant computer architecture needs to be developed. Fabricating reliable interconnects between devices using carbon nanotubes (or some other technology) is an additional challenge. A significant amount of work is ongoing in each of these areas. Even though experimental progress to date in this area has been substantial, it seems unlikely (as with quantum computing) that molecular computers could be developed within the next 15 years that would be relatively attractive (from a price/performance standpoint) compared with conventional electronic computers.

Broader Issues and Implications

Examining the potential for developing qualitatively different computational capabilities from different technology bases is a challenging exercise. The history of computing over the last 50 years has seen one major shift in technology base (from vacuum tubes to semiconductor transistors), with a corresponding shift not just in computational power but also in attitudes about the value of computers. Ideas of computers as simple machines for computation gave way to the use of computers for personal productivity with the advent of the microprocessor. As the power of these microprocessors has grown exponentially, they have also been seen more recently as a vehicle for new media and socialization.

The ramifications of future computing technologies will be determined principally by two factors: the conception, development, and adoption of new applications that require significantly more computational power; and the ability of technology to address these demands. New applications are always difficult to anticipate, but it is less challenging to foresee the likely consequences for diffusion of this technology. Past experience with personal computers and telecommunications has shown that these technologies diffuse more rapidly in the developed world than in the developing world. It is difficult to foresee an increase in the political or ethical barriers to computing technologies beyond those seen today, and these are rapidly vanishing.

On the remaining question of technology development, the odds-on favorite for the next 15 years remains traditional digital electronic computers based on semiconductor technology. Given the virtual certainty of continued progress in this area, it is hard to imagine a scenario in which a competing technology (quantum-switch-based computing, molecular computers, or something else) could offer a significant performance advantage at a competitive price. But the longer-term, traditionally elusive question in the period after 2015 is: How long will traditional silicon computing last? And when, if ever, will a competing technology become available and attractive? If an alternative computing technology becomes sufficiently attractive, the economic effects of technology substitution on the current semiconductor industry and adjacent industries must be considered. For example, major industry players may be faced with a choice between cannibalizing their existing market opportunities in favor of these new, future technologies, competing head-on with new players, or simply acquiring them. Most important, given the very different architectural approaches of these technologies and the classes of problems for which they are best suited, what will be the effect on future applications? The promises of nanotechnology may indeed become a reality in the period after 2015, but it will face these competitive challenges before its significance becomes global.

INTEGRATED MICROSYSTEMS AND MEMS

MEMS is less an application area in itself than a manufacturing or fabrication technique that enables other application areas. Many authors use MEMS as shorthand to imply a number of particular application areas. As it is used here, MEMS is a "top-down" fabrication technology that is especially useful for integrating mechanical and electrical systems together on the same chip. It is grouped in the category of integrated microsystems because these same MEMS techniques can be extended in the future to also help integrate biological and chemical components on the same chip, as discussed below. Thus far, MEMS techniques have been used to make some functional commercial devices such as sensors and single-chip measurement devices. Many researchers have used MEMS technologies as analytical tools in other areas of nanotechnology such as the ones discussed here.

Smart Systems-on-a-Chip (and Integration of Optical and Electronic Components)

Simple electro-optical and chemical sensor components have already been successfully integrated onto logic and memory chip designs in research and development labs. Likewise, radio frequency component integration in wireless devices is already being produced in mass quantities. Some companies have products capable of doing elementary DNA testing. The 1999 ITRS (noted above) predicts the introduction of chemical sensor components with logic in commercial designs by 2002, with electro-optical component integration by 2004, and biological systems integration by 2006. Given these predictions, there is clearly time for relatively complex integrated systems and applications to develop within the 2015 timeline. These advances could enable many applications where increased integrated functionality can become ubiquitous as a result of lower costs and micro-packaging.

Micro/Nanoscale Instrumentation and Measurement Technology

Instrumentation and measurement technologies are some of the most promising areas for near-term advancements and enabling effects. As optical, fluidic, chemical, and biological components can be integrated with electronic logic and memory components on the same chip at marginal cost, drug discovery, genetics research, chemical assays, and chemical synthesis are all likely to be substantially affected by these advances by 2015 (see also the previous section).

Some of the first applications of nanoscale (and microscale) instruments were as basic sensors for acceleration (such as those used in airbags), pressure, etc. Small, microscale, special-purpose optical and chemical sensors have been used for some time in sophisticated laboratory equipment, along with microprocessors for signal processing and computation. Already, companies have produced products that allow for basic DNA analysis, and that can assist in drug discovery. As these sensors become more sophisticated and more integrated with computational capability (with the aid of systems-on-a-chip), their utility should grow tremendously, especially in the biomedical arena.

Broader Issues and Implications

There are several advantages of nanotechnology for integrated systems in general (and instrumentation and measurement systems as a subset of these). First, existing semiconductor technologies will likely allow the volume manufacture of integrated smart systems that can be produced at low enough cost to be considered disposable. Second, the massive parallelism afforded by this same technology allows for the rapid analysis (with integrated computation) of very complex samples (such as DNA), the processing of large numbers of samples, and the recognition of large numbers of agents (e.g., infectious agents and toxins). Devices with these properties are already of tremendous utility in the biomedical arena for drug testing, chemical assays, etc. In addition, they will likely find utility in a variety of industrial applications.

Integrated micro/nanosystems are already starting to affect applications where miniaturization of components, subsystems, and even complete systems is significantly reducing device size, power, and consumables while introducing new capabilities. This area lends itself naturally to the confluence of all the broad areas discussed in this report (biotechnology, materials, and nanotechnology). The next five to ten years will likely see the integration of computational capabilities with biological, chemical, and optical components in systems-on-a-chip. At the same time, advances in biotechnology should drive applications for drug discovery and genomics, as well as the basic understanding of many other phenomena. Advances in biomaterials will likely produce biologically compatible packaging, capable of isolating substances from the body in a time-controlled fashion (e.g., for drug delivery). The confluence of these capabilities could allow for continued development of microscale and nanoscale systems that could continue to be introduced into the body to perform basic diagnostic functions in a minimally invasive way,[14] providing new abilities to remedy health problems.

Other possible applications include: pervasive, self-moving sensor systems; nanoscrubbers and nanocatalysts; even inexpensive, networked "nanosatellites." For example, so-called "nanosatellites" are targeting order-of-magnitude reductions in both size and mass (e.g., down to 10 kg) by reducing major system components using integrated microsystems. If successful, this could economize current missions and approaches (e.g., communication, remote sensing, global positioning, and scientific study) while enabling new missions (e.g., military tactical space support and logistics, distributed sparse aperture radar, and new scientific studies) (Luu and Martin, 1999 [214]). In addition, advances could empower the proliferation of currently controlled processing capabilities (e.g., nuclear isotope separation) with associated threats to international security. Progress will likely depend on investment levels as well as continued S&T development and progress.

MOLECULAR MANUFACTURING AND NANOROBOTS

Technology

A number of experts (K. Eric Drexler, among others) have put forth the concept of molecular manufacturing where objects are assembled atom by atom (or molecule by molecule).[15] Bottom-up molecular manufacturing differs from microtechnology and MEMS in that the latter employ top-down approaches using bulk materials using macroscopic fabrication techniques.

To realize molecular manufacturing, a number of technical accomplishments are necessary. First, suitable molecular building blocks must be found. These building blocks must be physically durable, chemically stable, easily manipulated, and (to a certain extent) functionally versatile. Several workers in the field have suggested the use of carbon-based diamond-like structures as building blocks for nano-mechanical devices, such as gears, pivots, and rotors. Other molecules could also be used to build structures, and to provide other integrated capabilities, such as chemically reactive structures. Much additional work in the area of modeling and synthesis of appropriate molecular structures is needed, and a number of groups are working to this end. Dresselhaus and others are fabricating suitable molecular building blocks for these structures.

The second major area for development is in the ability to assemble complex structures based on a particular design. A number of researchers have been working on different approaches to this issue. Different techniques for physical placements are under development. One approach by Quate, MacDonald, and Eigler uses atomic-force or molecular microscopes with very small nanoprobes to move atoms or molecules around with the aid of physical or chemical forces. An alternative ap-

[14]See, for example, recent advances in wireless capsule endoscopy (Iddan et al., 2000 [210]).

[15]See Drexler, 1987; Drexler, 1992; Nelson and Shipbaugh, 1995; Crandall, 1996; Timp, 1999; Voss, 1999; and Zachary, 2000 [162–168].

proach by Prentiss uses lasers to place molecules in a desired location. Chemical assembly techniques are being addressed by a number of groups, including Whiteside's approach to building structures one molecular layer at a time.

A third major area for development within molecular manufacturing is systems design and engineering. Extremely complex molecular systems at the macro scale will require substantial subsystem design, overall system design, and systems integration, much like complex manufactured systems of the present day. Although the design issues are likely to be largely separable at a subsystems level, the amount of computation required for design and validation is likely to be quite substantial. Performing checks on engineering constraints, such as defect tolerance, physical integrity, and chemical stability, will be required as well.

Some workers in the area have outlined a potential path for the evolution of molecular manufacturing capability, which is broken down by overall size, type of fabrication technology, system complexity, component materials used, etc. Some versions of this concept foresee the use of massively parallel nanorobots or scanning nanoprobes to assemble structures physically (with 100 to 10,000 molecular parts). Other, more advanced concepts incorporate chemical principles and use simple chemical feedstocks to achieve much larger devices on the order of 10^8 to 10^9 molecular parts.

Ostensibly, as each of these techniques matures (or fails to develop), more systems and engineering-level work must be done before applications can be realized on a significant scale. Although molecular manufacturing holds the promise of significant global changes (such as retraining large numbers of manufacturing workforces, opportunities for new regions to vie for dominance in a new manufacturing paradigm, or a shift to countries that do not have legacy manufacturing infrastructures), it remains the least concrete of the technologies discussed here. Significant progress has been made, however, in the development of component technologies within the first regime of molecular manufacturing, where objects might be constructed from simple molecules and manufactured in a short amount of time via parallel atomic force microprobes or from simple self-assembled structures. Although the building blocks for these systems currently exist only in isolation at the research stage, it is certainly reasonable to expect that an integrated capability could be developed over the next 15 years. Such a system could be able to assemble structures with between 100 and 10,000 components and total dimensions of perhaps tens of microns. A series of important breakthroughs could certainly cause progress in this area to develop much more rapidly, but it seems very unlikely that macro-scale objects could be constructed using molecular manufacturing within the 2015 timeframe.

Broader Issues and Implications

The present period in molecular manufacturing research is extremely exciting for a number of reasons. First, many workers have begun to experimentally demonstrate basic capabilities in each of the core areas outlined above. Second, continued progress and ongoing challenges in the area of top-down microelectronics manufacturing are pushing existing capabilities closer to the nanoscale regime. Third, the understanding of fundamental properties of structures at the nanoscale has been greatly enhanced by the ability to fabricate very small test objects, analyze them experimentally using new capabilities, and understand them more fundamentally with the aid of sophisticated computational models.

At the same time, many visionaries have advanced notions about potential applications for molecular manufacturing. But because experimental capabilities are in their infancy (as many workers have pointed out), it is extremely difficult to foresee many outcomes, let alone assess their likelihood.

International competition for dominance or even capability in cutting-edge nanotechnology may still remain strong, but current investments and direction indicate that the United States and Europe may retain leadership in most of this field.[16] Progress in nanotechnology will depend heavily on R&D investments; countries that continue to invest in nanotechnology today may lead the field in 2015. In 1997, annual global investments in nanotechnology were as follows: Japan at $120 million, the United States at $116 million, Western Europe at $128 million, and all other countries (former Soviet Union, China, Canada, Australia, Korea, Taiwan, and Singapore) at $70 million combined (Siegel et al., 1999 [163]). Funding under the U.S. National Nanotechnology Initiative is proposed to increase to $270 million and $495 million in 2000 and 2001, respectively (National Nanotechnology Initiative, 2000 [179]).

This would not preclude other countries from acquiring capabilities in nanotechnology or in using these capabilities for narrow technological surprise or military means. Given the difficulty in foreseeing outcomes and estimating likelihoods, however, it is also difficult to extrapolate predictions of specific threats and risks from current trends.

---

[16]See also the longer discussion of international competition in the discussion of meta-technology trends in Chapter Three.

DISCUSSION

THE RANGE OF POSSIBILITIES BY 2015

Impossible though it is to predict the future, technology trends give some indication of what we might anticipate based on current movements and progress. As discussed, the progress and effect of these trends will be modulated by enablers and barriers. Furthermore, these trends could have various effects on the world. Figures 3.1, 3.2, and 3.3 tie these components together for three trends: genetically modified foods, smart materials, and nanotechnology.

Figure 3.1 shows the range of potential paths that genetically modified foods might take by 2015 along with enablers, barriers, and effects. Investments and genome decoding are fueling the ability to modify and engineer organisms to provide needed capabilities, but social concerns are already affecting the generation and use of GM foods, especially between the United States and European Union (particularly in the United Kingdom). In an optimistic 2015, GM foods will be widespread, resulting in significant benefits for food quality, global production, and the environment (e.g., represented by the Biotechnology Industry Organization's positions (BIO, 2000 [41]). Policy controls or lack of investments might moderate the production and use of GM foods, leading to increased reliance on traditional mechanisms for food productivity increases and pest control.

Figure 3.2 shows the range of potential paths that smart materials might take by 2015 along with enablers, barriers, and effects. Investments and commitment to research are prime enablers, but limited funding, limited labor, failing interests, or lack of public acceptance of highly monitored environments could modulate growth and application. In an optimistic 2015, smart materials could be used in a wide array of novel applications. Barriers, however, could slow the development and application of smart materials to, say, advanced sensors with integrated actuator capabilities.

Figure 3.3 shows a striking range of opinions on where nanotechnology might be by 2015 along with enablers, barriers, and effects. Current high-visibility investments and technology breakthroughs will be needed to realize the full potential of nanotechnology, but research and development costs, applicability, complexity, accessibility, and even social acceptance (e.g., of intelligent nanomachines) could slow its growth. The optimistic future state is perhaps best exemplified by the vision of pervasive nanotechnology involving molecular manufacturing of a host of nanosystems

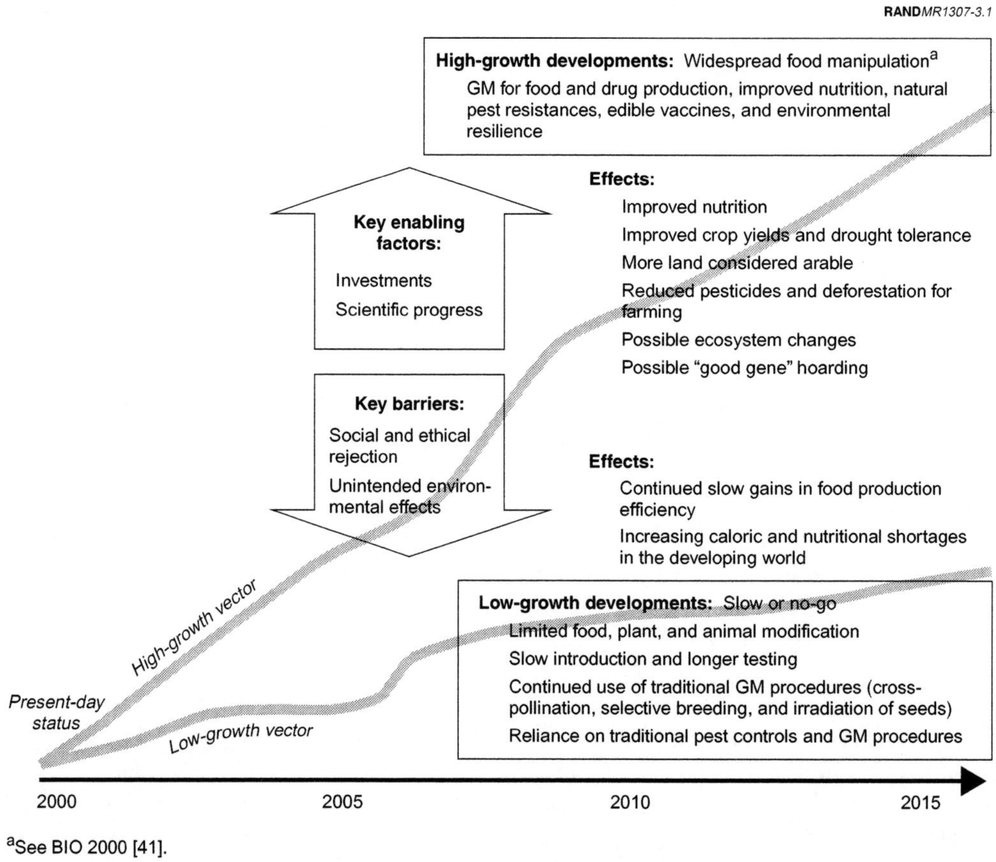

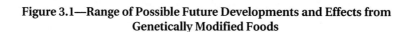

Figure 3.1—Range of Possible Future Developments and Effects from Genetically Modified Foods

with revolutionary capabilities (see Drexler, 1987, 1992 [162, 163]); moreover, nanomanufacturing would take place on a global scale, giving developing countries the opportunity to invest in and participate in the revolution. From a more pragmatic view, lack of technological breakthroughs might limit the results by 2015 to an evolutionary path where the current trend to smaller, faster, and cheaper systems continues through nano-level advances in semiconductor production to continue Moore's Law (see SEMATECH, 1999 [190]).

Table 3.1 shows the facilitative relationships of four technologies along with their individual high-growth futures, low-growth futures, effects, enablers, and barriers. These relationships emphasize that well-know technologies such as information technology and biotechnology actually rely on less-known enabling technologies for some of their progress. Although these facilitative relationships impart dependency on other technologies, the combined effect will accelerate the capability and promise of technology as long as key enablers can be maintained.

RANDMR1307-3.2

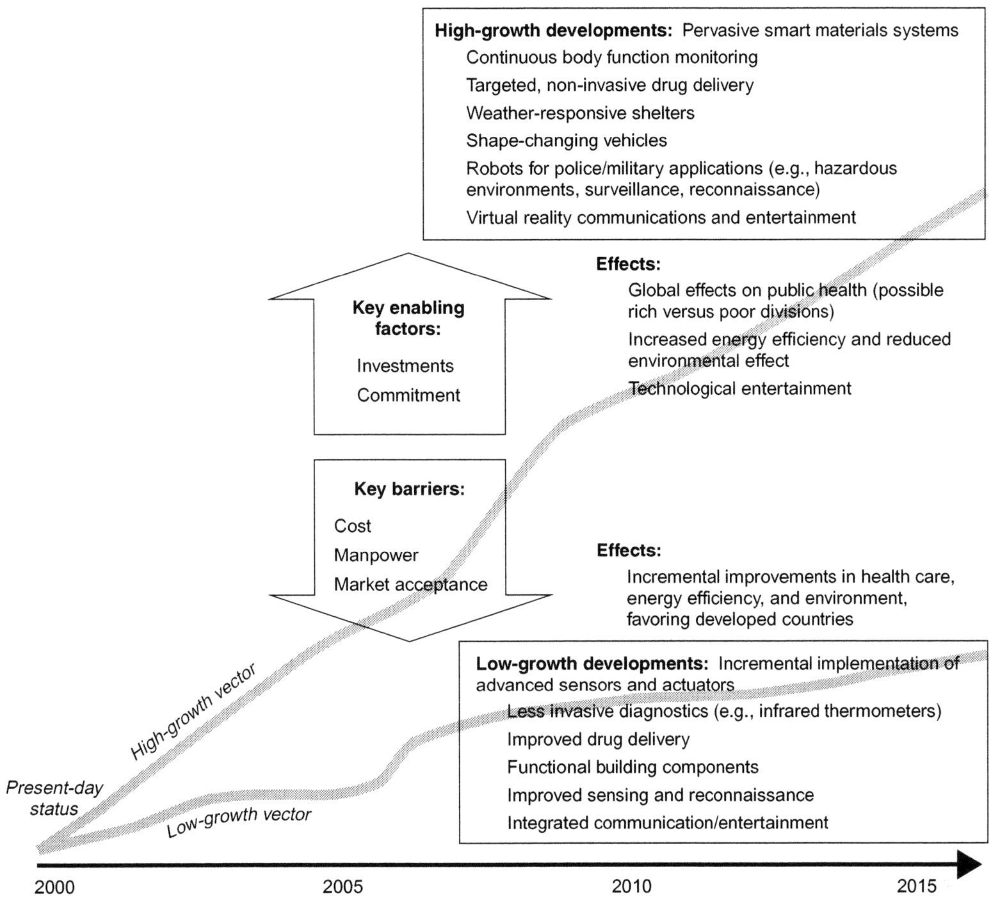

Figure 3.2—Range of Possible Future Developments and Effects of Smart Materials

META-TECHNOLOGY TRENDS

A number of meta-trends can be observed by reviewing the technology trends discussed above and the discussions in the open literature. These meta-trends include the increasingly multidisciplinary nature of technology, the accelerating pace of change and concerns, increasing educational demands, increased life spans, the potential for reduced privacy, continued globalization, and the effects of international competition on technology development.

Multidisciplinary Nature of Technology

Many technology trends have been enabled by the contributions of two or more intersecting technologies. Consider, for example, MEMS-based molecular diagnostics, biomaterials, biological-based computing, and biomimetic robotics. Various technologies have combined in the past to enable applications, but there has been an increase in multidisciplinary teaming to examine system challenges and envision

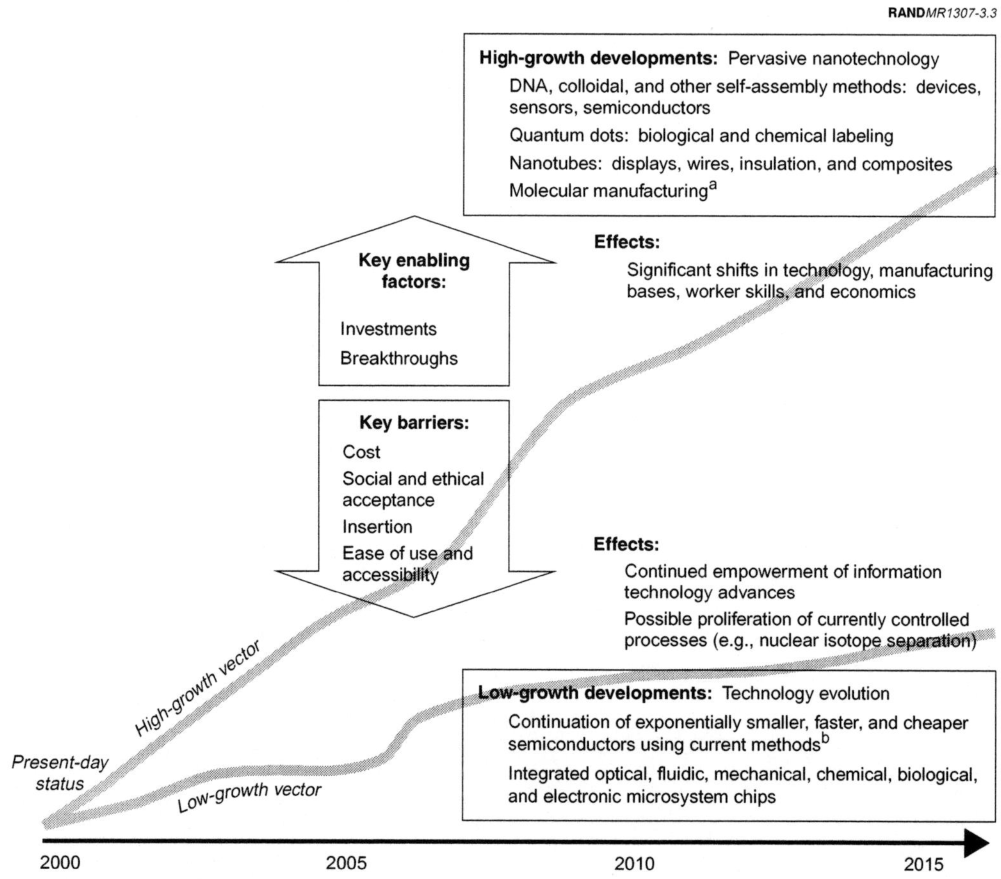

[a]See Drexler, 1987, 1992 [162, 163].
[b]See SEMATECH, 1999 [190].

Figure 3.3—Range of Possible Future Developments and Effects of Nanotechnology

approaches in a unified way rather than through a hierarchical relationship. Materials scientists, for example, are working increasingly with computer scientists and application engineers to develop biomedical materials for artificial tissues or to develop reactive materials to facilitate active system control surfaces. Materials are also being developed and adapted as embedded sensors and actuators for smart structures.

Figure 3.4 illustrates examples of how nanotechnology (scale), information technology (processing), materials (processing and function), and living organisms interrelate to produce new systems and concepts. Materials provide function, and the emergence of nanotechnology has enabled construction at a scale that integrates function with processing (smart materials). The combination of materials technology with biology and the life sciences has at the same time provided knowledge and materials obtained from living organisms to enable a further integration with pervasive effects from design (biomimetics) to end product (bionics). Note that these examples show different combinations of technologies resulting in different advances;

Table 3.1

The Range of Some Potential Interacting Areas and Effects of the Technology Revolution by 2015

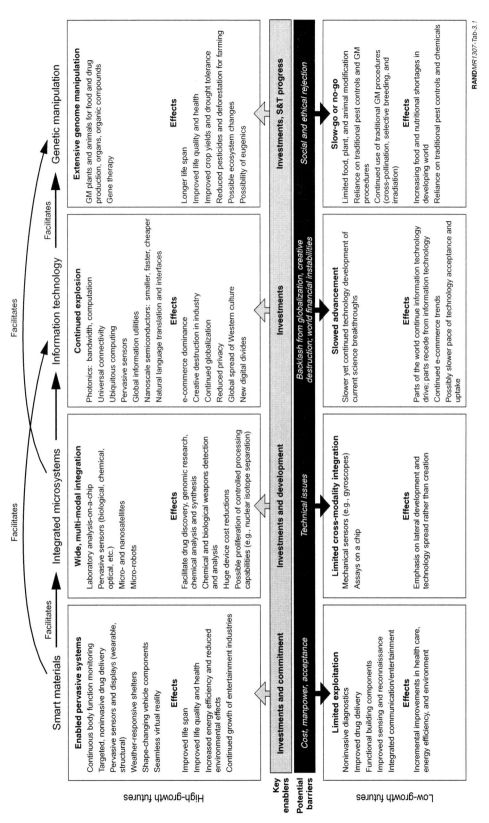

Smart materials —Facilitates→ Integrated microsystems → Information technology —Facilitates→ Genetic manipulation; Smart materials —Facilitates→ Information technology; Integrated microsystems —Facilitates→ Genetic manipulation

	Smart materials	Integrated microsystems	Information technology	Genetic manipulation
High-growth futures	**Enabled pervasive systems** Continuous body function monitoring Targeted, noninvasive drug delivery Pervasive sensors and displays (wearable, structural) Weather-responsive shelters Shape-changing vehicle components Seamless virtual reality **Effects** Improved life span Improved life quality and health Increased energy efficiency and reduced environmental effects Continued growth of entertainment industries	**Wide, multi-modal integration** Laboratory analysis-on-a-chip Pervasive sensors (biological, chemical, optical, etc.) Micro- and nanosatellites Micro-robots **Effects** Facilitate drug discovery, genomic research, chemical analysis and synthesis Chemical and biological weapons detection and analysis Huge device cost reductions Possible proliferation of controlled processing capabilities (e.g., nuclear isotope separation)	**Continued explosion** Photonics: bandwidth, computation Universal connectivity Ubiquitous computing Pervasive sensors Global information utilities Nanoscale semiconductors: smaller, faster, cheaper Natural language translation and interfaces **Effects** e-commerce dominance Creative destruction in industry Continued globalization Reduced privacy Global spread of Western culture New digital divides	**Extensive genome manipulation** GM plants and animals for food and drug production, organs, organic compounds Gene therapy **Effects** Longer life span Improved life quality and health Improved crop yields and drought tolerance Reduced pesticides and deforestation for farming Possible ecosystem changes Possibility of eugenics
Key enablers	**Investments and commitment**	**Investments and development**	**Investments**	**Investments, S&T progress**
Potential barriers	*Cost, manpower, acceptance*	*Technical issues*	*Backlash from globalization, creative destruction; world financial instabilities*	*Social and ethical rejection*
Low-growth futures	**Limited exploitation** Noninvasive diagnostics Improved drug delivery Functional building components Improved sensing and reconnaissance Integrated communication/entertainment **Effects** Incremental improvements in health care, energy efficiency, and environment	**Limited cross-modality integration** Mechanical sensors (e.g., gyroscopes) Assays on a chip **Effects** Emphasis on lateral development and technology spread rather than creation	**Slowed advancement** Slower yet continued technology development of current science breakthroughs **Effects** Parts of the world continue information technology drive; parts recede from information technology Continued e-commerce trends Possibly slower pace of technology acceptance and uptake	**Slow-go or no-go** Limited food, plant, and animal modification Reliance on traditional pest controls and GM procedures Continued use of traditional GM procedures (cross-pollination, selective breeding, and irradiation) **Effects** Increasing food and nutritional shortages in developing world Reliance on traditional pest controls and chemicals

RAND*MR1307-Tab-3.1*

RANDMR1307-3.4

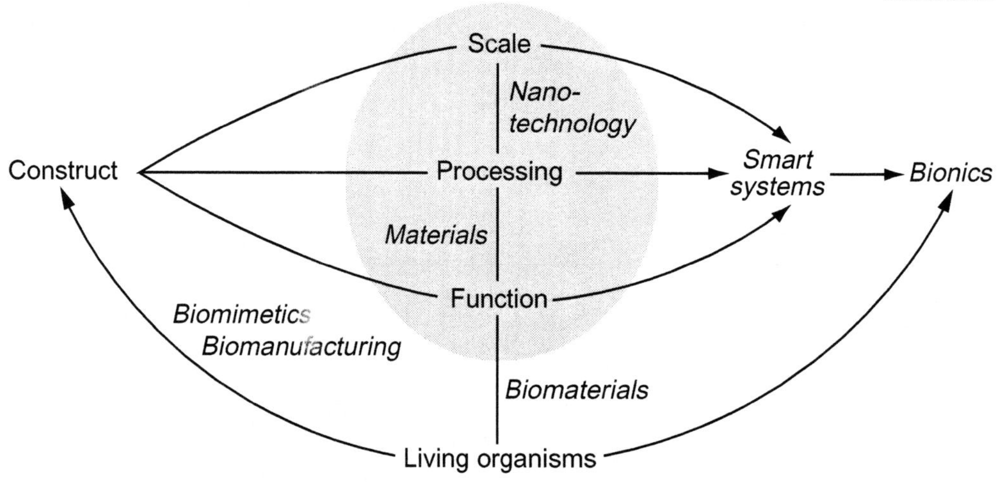

Figure 3.4—The Synergistic Interplay of Technologies

the entire set of technologies provides a rich mix of contributions to the overall technology revolution.

In addition to their technology and artifacts, different fields also tend to produce different views and approaches to the world. Combining these views also enriches the scientific toolbox used on a problem, resulting in advances that combine the best of each world and enabling applications that would not be possible otherwise. For example, engineers increasingly turn to biologists to understand how living organisms solve problems in the natural environment. Rather than blindly copying nature, such "biomimetic" endeavors often combine the best solutions from nature with artificially engineered components to develop a system that is better for the particular environment than any existing organism.

Many significant trends leverage technologies from multiple areas. Figure 3.4 shows examples of the interplay between biotechnology, nanotechnology, materials technology, and information technology areas. Smart materials are contributing both function and processing capabilities. In this figure we show how smart systems are enabled by progress in materials (sensor/actuator) and information technology capabilities together with microsystem trends. Smart systems in turn can be engineered to provide living organisms with bionic capabilities. Living organisms are informing new ways to construct systems (biomimetics) as well as manufacturing components themselves.

Accelerating Pace of Change

The general pace of technological advance and change seems to be accelerating. Economic growth, especially in the United States, is fueling applied research and development investments, resulting in new product innovations and approaches. Computer technology continues to advance to the point where products become obsolete in two to three years. In some areas of biomedical engineering the pace is even faster; some medical devices are obsolete by the time a prototype is developed (Grundfest, 2000 [107]). Such a pace could make it more difficult for legal and ethical advances to keep up with technology.

Accelerating Social and Ethical Concerns

As new technologies enable greater ability to manipulate the environment and living things, societal and ethical concerns are accelerating. Privacy, intellectual property, and environmental sustainability issues are all raised as new capabilities that are offered by technology.

Increased Need for Educational Breadth and Depth

Combined with an increased pace of technological change will likely come an increased need for continued learning and education. Just as computer skills are becoming more important today, both blue-collar and white-collar workers will likely need to improve their skills in other areas to avoid obsolescence in technological realms.

The multidisciplinary nature of technology is also changing the skills required by the workforce as well as R&D technologists. Developers increasingly need to understand vocabulary and fundamental concepts from other fields to work effectively in multidisciplinary teams, demanding more time in breadth courses. This trend may increase over time to the point where multidisciplinary degrees may be necessary, especially for visionaries and researchers who tie concepts together.

Finally, the population as a whole will likely need to have a wider understanding of science and technology to make informed political and consumer decisions. For example, current controversies regarding genetically modified foods require an open and questioning mind to be able to balance the often-complicated arguments made by various parties in the debate. Likewise, understanding the privacy implications and potential gains of heavily instrumented and monitored homes is needed to have an informed electorate and consumer base.

Longer Life Spans

Health-related advances hold the promise of continuing the trend of increasing human life spans in the developed world. This trend raises issues related to increased population, care for the elderly, and retirement living. Medical advances may also increase the quality of life, enabling people to not only live longer but to remain productive members of society longer.

Reduced Privacy

Various threats to individual privacy include pervasive sensors, DNA "fingerprinting," genetic profiles that indicate disease predispositions, Internet-accessible databases of personal information, and other information technology threats.

Privacy issues will likely result in legislative debates concerning legal protections and regulations, continued social and ethical debates about technology uses, the generation of privacy requirements and markets, and privacy-supporting technologies (e.g., security measures and components in sensory and information architectures and components). The timeliness, pervasiveness, and rationality of privacy concerns may dictate whether privacy issues are addressed in proactive or reactive ways. In recent history, however, privacy and security have taken a back seat to functionality and performance.[1] It is unlikely that privacy concerns will halt these technology trends, resulting in reducing privacy across the globe in measure with the amount of technology in a region. Scrutiny of privacy issues, however, may change public behavior in how it uses technology and may influence technological development by highlighting privacy as a social demand.

Continued Globalization

Globalization is likely to be facilitated not only by advances in information technology, the Internet, communications, and improved transportation (see, for example, Friedman, 2000 [217]) but also by enabled trends such as agile manufacturing where local investments in infrastructure could enable new players to participate in global manufacturing.

International Competition

Regarding international competition for developing cutting-edge technology, a range of possibilities exists in each area. These possibilities range from a nationally competitive system in which both technology investments and technology products are stovepiped with respect to national boundaries, to a situation in which they are highly fluid across national and regional boundaries. The actual direction will depend on a number of factors, including future regional economic arrangements (e.g., the European Union), international intellectual property rights and protections, the character of future multi-national corporations, and the role and amount of public sector research and development investments. Currently, there are moves toward competition among regional (as opposed to national) economic alliances, increased support for a global intellectual property protection regime, more globalization, and a division of responsibilities for R&D funding (public sector research funding with private sector development funding).[2] Naturally, these meta-trends are subject to change in accordance with the factors outlined elsewhere in this report.

---

[1]For example, security and privacy on personal computers and the Internet have been an afterthought in many cases. The marketplace has mostly ignored these issues until actual incidents and damages have forced the issue, raising concern and market demands.

[2]Note that even though private sector R&D expenditures are currently increasing in absolute dollars, many of these investments are relegated to relatively expensive development efforts instead of research.

CROSS-FACILITATION OF TECHNOLOGY EFFECTS

Beyond individual technology effects, the simultaneous progress of multiple technologies and applications could result in additive or even synergistic effects. Table 3.2 shows the results of an exercise where pairings of sample technology innovations were examined for such effects. Some advances will introduce capabilities that could be used to aid other advances and hence accentuate their effect beyond what would be achievable if the effects were independent and merely additive. It is also possible that certain combinations of realized advances could have negative effects on each other, resulting in unforeseen difficulties. Unforeseen ethical, public concern, or environmental difficulties may be examples.

Nine potential innovations were selected across biotechnology, materials technology, and nanotechnology to explore how technologies may facilitate each other. GM foods include the customization of crops and animals to improve nutrition and production while reducing pesticide and water use. Drug-testing simulations will improve drug development by simulating drug-body interactions to improve testing and understanding of drug interactions and population problems. Minimally invasive surgery (along with artificial tissues, structures, organs, and prosthetics) will improve health by addressing medical problems with reduced intervention and thus reducing cost and time while improving efficacy. Artificial heart tissue will reduce heart problems by providing regenerative materials to repair hearts. Personal identification databases will develop device materials to facilitate the protected (off-line) storage of information on an individual or in a small, portable system (e.g., a next-generation smart card). The global business enterprise will use rapid prototyping and agile manufacturing to leverage global production capabilities. A micro-locator tag will combine wireless communications at longer distances than current tagging technology and become commonplace in businesses and homes to facilitate logistics, the location of items, and interfaces with information processors (e.g., to control manufacturing parts or to plan meals based on available items in a pantry). An *in vivo* nanoscope would provide wireless, in-body testing and monitoring of medical conditions, replacing wired probes and measuring factors impossible with today's technology. Finally, cheap catalytic air "nanoscrubber" (a molecular manufacturing "wild card") would be produced in massive quantities and released into the atmosphere to convert carbon molecules to less harmful forms to decrease the environmental effects of fossil fuels.

Each innovation was rated as either likely (circle) or uncertain (square) or in-between (circle–square). Major effects of each innovation were also listed and rated using these tags. Additional details of these effects are included in the shaded boxes on the diagonal. The degree of shading of these boxes along the diagonal denotes the potential scope of the innovation (global or moderated) by 2015. Here *global* (cross-hatched gray boxes) denotes widespread effects; *moderated* (medium gray boxes)

Table 3.2

Potential Technology Synergistic Effects

		Biotechnology: Genetically Modified Foods	Biotechnology: Drug Testing Simulations	Biotechnology: Minimally Invasive Surgery
Biotechnology	**Genetics** ● • Genetically Modified Foods □ – Develop customized foods/ types for different climates	*↑ Food production* *↑ Nutrition* *↓ Environmental effect*	**↑ Health**	**↑ Health**
Biotechnology	**Computational biology** □ • Drug Testing Simulations ● – Biopharma shifts: rescued drugs; shift to custom drugs and diagnosis		*Industry* *↑ Dev. ↓$ ↓Time* *Custom drugs? ↑*	**↑ Health**
Biotechnology	**Biomedical engineering** ●–□ • Minimally Invasive Surgery, artificial tissues/structures/ organ, neural prosthetics ● – Health/life expectancy/cost			*↓ Cost ↓Time $* *↑ Life expectancy* *→ Social: retirement demographics*
Materials technology	**Tissue engineering** ● • Artificial Heart Tissue ● – Treat heart attack with regenerated tissue			
Materials technology	**Smart materials** □ • Personal ID/Database ●–□ – Instant, secure ID/data			
Materials technology	**Agile manufacturing** ●–□ • Global Business Enterprise (consumer— direct order → mfg. to order → deliver, maintain, track) ● – Power of business NGOs ↑			
Nanotechnology	**Smart system-on-a-chip** ● • Micro-Locator Tag with communications ●–□ – Enables persistent surveillance and logistics efficiency			
Nanotechnology	**Nano-instrumentation** □ • *In vivo* Nanoscope (bio-measurements/genetics) ● – Timely health information			
Nanotechnology	**Molecular manufacturing** □ • Catalytic Air Nanoscrubber: Molecular-scale for removal of CO, CO_2 at source □ – Vast decrease in environmental effect of fossil-fuel consumption			

Legend: □ Uncertain; ● Likely; (hatched) Global; (shaded) Moderated; (—) No additive or synergistic effect

Shading: Global — Genetics row in all three columns. Moderated — Computational biology row under Drug Testing Simulations and Minimally Invasive Surgery; Biomedical engineering row under Minimally Invasive Surgery.

RANDMR1307-Tab-A

indicates that the effects will likely be constrained across some dimension (e.g., geographic, industrial scope, economic access) in the 2015 timeframe. For example, if they survive public concern, GM foods could become pervasive across the globe, affecting most agriculture. The effects of drug-testing simulation, however, may be constrained to financial gains within the pharmaceutical industry or within the developed world that can afford a new wave of customized drugs. Note that these trends are moving toward globalization, but moderation across some dimension may indicate that their effects beyond 2015 may be through trickle-down adoption as costs become more acceptable to more populations.

The rest of the cross entries in Table 3.2 indicate whether potential synergistic effects may occur if both innovations come to pass.

Table 3.2

(continued)

Materials technology			Nanotechnology		
Artificial Heart Tissue	Personal ID/Database	Global Business Enterprise	Micro-Locator Tag	*In Vivo* Nanoscope	Catalytic Air Nanoscrubber
↑ Health	—	**New distribution question**	—	**Health needs assessment** **Govt. ID of GMO** **↑ Lab instrument**	**↑ Health expectancy** **? Energy effect**
↑ Health **↑ Drugs to maintain tissue**	**↑ Health data on genome and drugs**	—			—
(Subset)	**(Subset)**			**Facilitation**	—
↓ *Death rate, eliminate premature death from congenital problems*	—			**Facilitation**	—
	Instant remote purchasing *Privacy barrier*			**Facilitation**	—
		↑ *Consumer power* ↓ *Government control (effect on international trade)*		—	**Facilitation**
			↑ *Industrial efficiency* *Privacy barrier*		
				↑ *Health benefits/ preventative medicine*	

☐ Uncertain ● Likely ▨ Global ▭ Moderated ⊟ No additive or synergistic effect

RAND*MR1307-Tab-B*

- Some cross effects may be *additive* in the same dimension. For example, both GM foods and tissue engineering could increase health benefits.
- Other cross effects may not be merely additive but may *facilitate* each other, enabling new capabilities or increasing their individual effects. For example, *in vivo* nanoscopes could facilitate the benefits of biomedical engineering and tissue engineering by improving our ability to diagnose and apply the correct engineered remediation for individual patients.
- Finally, some cross effects may be *mutually exclusive* and result in no additive or synergistic effect; these are indicated with gray boxes containing dashes For example, a personal ID system could have effects that are largely independent of the effects of GM foods.

These observations should not be viewed as predictions of the future state by 2015 but rather as an effort to examine the potential scope of the effects of technology trends, including assessments of interactive effects if pairs of innovations come to fruition. The cross effects were assumed to be symmetric; thus, only one-half of the table was shown.

THE HIGHLY INTERACTIVE NATURE OF TREND EFFECTS

The effects (e.g., social, economic, political, public opinion, environmental) of technology trends often interact with each other and result in subsequent effects. Figure 3.5 illustrates this interactive nature for a trend that has already entered public debate and thus is already having global effects. In this example, we show how increased population (and thus demand for food productivity increases) is a major driver for the use of GM foods. We also show how subsequent effects can contradict each other and drive policy decisions.

Such an influence chart can therefore be useful in following the logical arguments made by multiple individuals and organizations on a topic of debate and to understand how the points made fit into a larger picture of interactions. For example, the genetic modification safety issue appears in the context of three fundamental players: companies (searching for new markets and pushing for patenting rights), anti-GM activists (trying to eliminate genetic modification altogether), activists for developing countries and world food supplies (working to improve and tailor crops), and environmentalists (worrying about biodiversity as well as deforestation). These interacting drivers sometimes conflict with each other and sometimes facilitate each other. It is unclear what will happen politically. A compromise point may be reached that balances intellectual property protections with developing-world market needs. Technology and education may address many safety concerns while enabling continued use of GMOs. Environmentalists may find a balance where extensive customization of crops and reduced deforestation may address biodiversity concerns. It is unclear which position will prevail politically or even to what extent such technology can be regulated, but constructing such charts and following the progression of the interacting debates can help to monitor the situation and inform policy.

Many of the effects discussed in the technology section above are foresights of what may happen as a result of the trends discussed, but because the effects will be felt so far in the future, it is often hard to understand what the interactive and subsequent effects may be. This does not mean that the final effects will not be complex. Rather, the reader must be aware of the complex nature of the effects and continue to look for them as the trends and technologies mature.

Figure 3.5 demonstrates how potential technology effects interact and intertwine between economic, political, public opinion, and environmental domains. Conflicts between effects are marked with an exclamation point (!) and show how the net effect could be balanced by a number of factors or policies. This figure is not complete but illustrates the complex interactions between technology trends. These effects

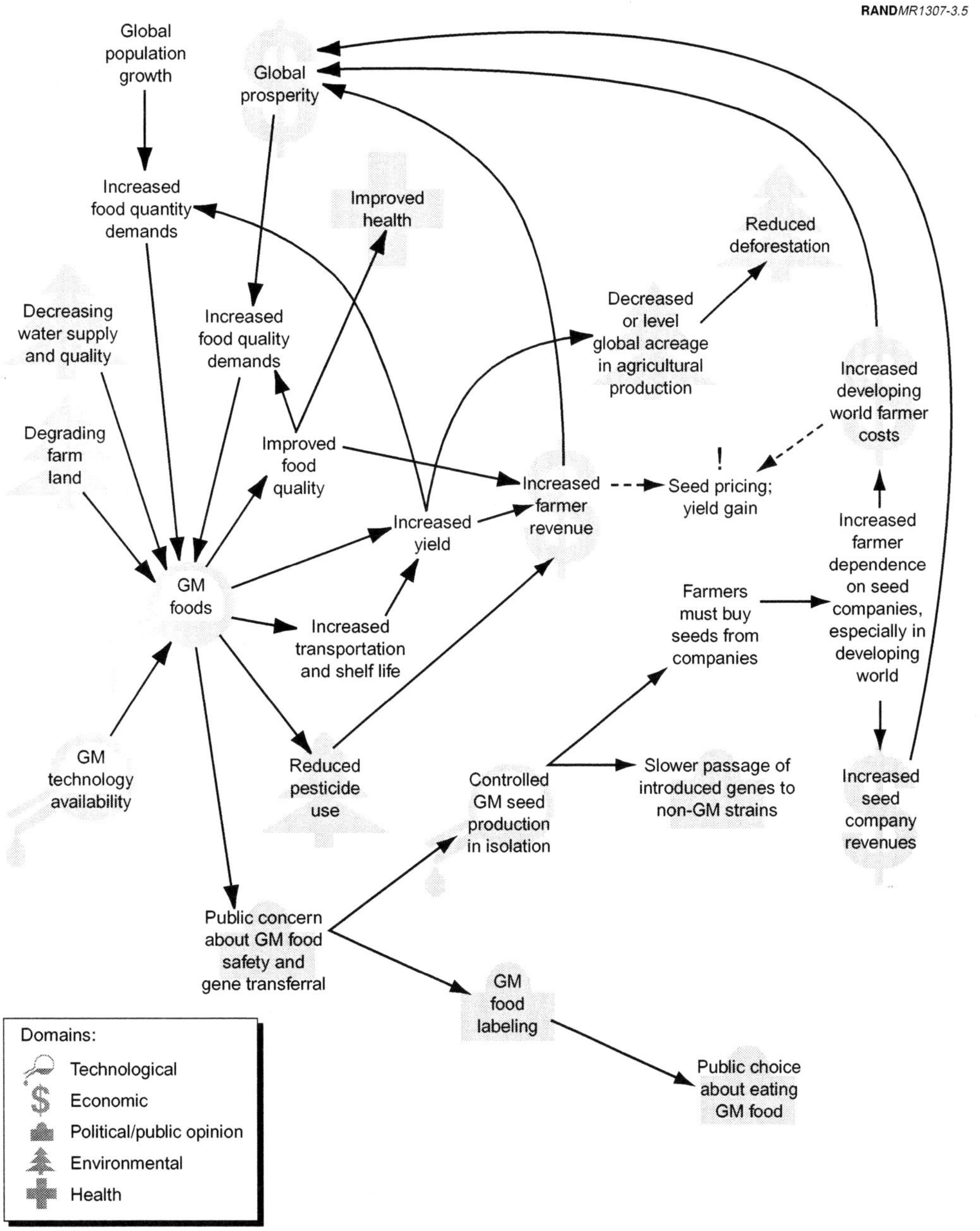

Figure 3.5—Interacting Effects of GM Foods

may be conditionalized to particular regions or conditions. For example, reduced pesticide use could mostly affect farmers who already use pesticides, but farmers who use GM crops with systemic pesticides could still reap increased yields and revenues.

THE TECHNOLOGY REVOLUTION

Beyond the agricultural, industrial, and information revolutions of the past, a multidisciplinary *technology revolution*, therefore, appears to be taking place in which the synergy and mutual benefit among technologies are enabling large advances and new applications and concepts (see Table 3.3). Many individual technology trends are pursuing general directions as shown. Beyond specific technologies, however, meta-trends are appearing that characterize properties of the technology trends and provide an abstract framework for describing the technology revolution. Furthermore, entry costs ("tickets") illustrate what individuals, businesses, countries, and regions will likely need to enter and continue to participate in the technology revolution.

Beyond individual technology trends and meta-trends, the prerequisites and resources required to participate in the technology revolution seem to be evolving.

Table 3.3

The Technology Revolution: Trend Paths, Meta-Trends, and "Tickets"

Past Technology	Present Technology	Future Technology
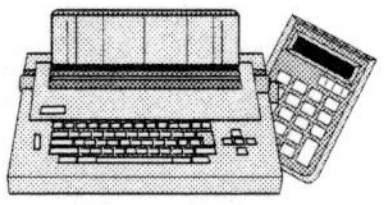		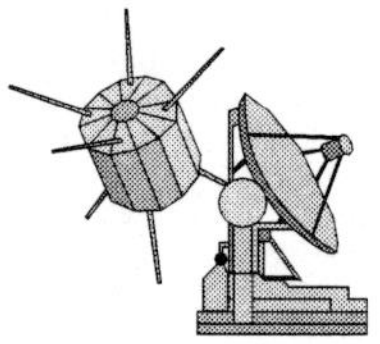
	Trend Paths	
Metals and traditional ceramics	Composites and polymers	Smart materials
Engineering and biology separate	Biomaterials	Bio/genetic engineering
Selective breeding	Genetic insertion	Genetic engineering
Small-scale integration	Very-large-scale integration	Ultra/giga-scale integration
Micron plus lithography	Sub-micron lithography	Nano-assembly
Main frame	Personal computer	Micro-appliances
Stand-alone computers	Internet-connected machines	Appliance and assistant networks
	Meta-Trends	
Single disciplinary	Dual/hierarchically disciplinary	Multi-disciplinary
Macro-systems	Micro-systems	Nano-systems
Local	Regional	Global
Physical	Information	Knowledge
	"Tickets" to the Technology Revolution	
Trade schools	Highly specialized training	Multidisciplinary training
General college	Specialized degree	Multidisciplinary degree(s)
Locally resourced products	Locally resourced components	Products tailored to local resources
Capital ($)	Increased capital ($$)	Mixed

RAND*MR1307-Tab-3.3*

The overall workforce will likely have to contribute to and understand an increasingly interdisciplinary activity. Just as computer skills are becoming more important today, a basic capability to work with or use new materials and processes involving biology and micro/nanosystems will likely be required. Not only will new skills and tools be needed, but we could see by 2015 a fundamental paradigm shift in the way we work and live because of the technology revolution.

Consumers and citizens should gain a basic understanding of technology to make informed decisions and demands on our political, social, economic, legal, and military systems. Likewise, scientists, engineers, technologists, and the government will have increasing responsibilities to think about and communicate the benefits and risks of technological innovations. Such knowledge does not need to be deep in each area, but a basic understanding will enable proper development and use of technology.

Technology workers (e.g., researchers, developers, and application designers) will likely need a deeper multidisciplinary education to enable teaming and to understand when to bring in specialists from different disciplines. Distance learning could facilitate the rapid dissemination of knowledge from developing specialists.

In addition to formal breadth courses and multidisciplinary training, the Internet may also facilitate the ability of people to acquire new knowledge in multiple disciplines and to keep skills current with developing trends. Authentication of both knowledge sources and training will remain important, especially for worker training, but demonstrated experience could continue to substitute for formal training.

Some of the progress in technology trends is enabled by multidisciplinary R&D teams. The old paradigm of hierarchical relations of technology is being replaced with one where a team searches for solutions in multiple disciplines. For example, materials are not relegated to providing infrastructure alone for traditional computing approaches but are being considered for processing applications themselves when smart materials can provide sensing and processing directly.

The use of and dependence on resources also seems to be evolving. In the past, local resources strongly influenced local production. Transportation currently allows local resources (e.g., natural resources or labor) to be combined with (value-added) resources from other areas, ultimately resulting in products that meet specific end-product needs. By 2015, end products might be tailored to utilize available resources and enable a wider range of technology participants.

Although current capital costs have been increasing for technology participants, it is unclear where this trend will lead by 2015. On the one hand, certain manufacturing and research equipment (e.g., for semiconductor fabrication) will likely continue to be more costly and be concentrated in the hands of a few manufacturers. On the other hand, genomic processing and rapid prototyping might be pursued with relatively low-cost equipment and with little infrastructure, allowing biological and part manufacturing practically anywhere in the world. Knowledge itself will become increasingly important and valuable. Generation, validation, and search for specific new knowledge in new technological domains could become increasingly costly with the increased availability of raw data (e.g., understanding the function and implications of genome maps). Such knowledge could become increasingly protected, yet global knowledge availability and transfer of public data and knowledge will likely be facilitated by information technology.

Other questions regarding participation make it unclear what will happen by 2015. Can global connectivity and distance learning make initial and continuing education and training globally available? Can they help bridge the gap between academic disciplines? Can agile manufacturing make it possible to participate in global manufacturing with *less* capital by producing components for larger products? Can advances in technology enable the tailored use of local resources more effectively?

THE TECHNOLOGY REVOLUTION AND CULTURE

The technology revolution is going far beyond merely generating products and services. First, these products and services are changing the way people interact and live. Cell phones are already bringing business and personal interactions into previously private venues. Increased miniaturization and sensorization of items such as appliances, clothing, property, and automobiles will likely change the way these devices interact with our lifestyles. The foods we eat are likely to be increasingly engineered. Health care could be integrated into our lives through better prognostics and daily monitoring for conditions.

Second, business is becoming increasing global and interconnected. This trend will likely continue, for example, with the aid of agile manufacturing and rapid prototyping.

Third, the requirements for participating in the generation of products and services are changing (see the bottom of Table 3.3). As technology becomes more interdisciplinary, education and training must change to enable workers to participate. Education should emphasize a larger component of breadth across disciplines to give at least a fundamental understanding of multiple disciplines. Businesses will likely need to spend more resources on continued training across their workforce.

Taken together, these trends indicate that technology is having a *cultural* effect. Modes of social interaction are changing. Both ideas and norms are influenced by newly introduced standards and the wider access to other cultural approaches.

Communities are already reacting to the cultural invasion in information technology (Hundley et al., [212]). Some cultures are very open to adapting new technology (especially given financial motivations); others are concerned that their cultural traditions are in danger of being replaced by a global (sometimes Western or American) cultural invasion and are less open to adopting and accepting technology. As trends enabled by biotechnology, materials technology, and nanotechnology expand the effect of the technology revolution, we anticipate that communities could continue to respond to the technology revolution in various ways.

As the pace of these changes is likely to be rapid during the next 15 years, these community responses to technology and its effect on local culture may result in increased conflict. Some conflict may be overt as communities and governments establish policies to protect extant culture[3] or even attempt to reject the technology revolution by various means. Other conflicts may be covert as individuals who reject technology turn to terrorism or technology attacks in an attempt to influence the change.

On the other hand, improvements in the quality of life resulting from the technology revolution could reduce conflict. Policies to enable the sharing of benefits may help to tilt the future toward this more positive outcome.

CONCLUSIONS

Beyond the agricultural and industrial revolutions of the past, a broad, multidisciplinary *technology revolution* is changing the world. Information technology is already revolutionizing our lives and will continue to be aided by breakthroughs in materials and nanotechnology. Biotechnology will revolutionize living organisms. Materials and nanotechnology are developing new devices with unforeseen capabilities. These technologies are affecting our lives. They are heavily intertwined, making the technology revolution highly multidisciplinary and accelerating progress in each area.

The revolutionary effects of biotechnology may be the most startling. Collective breakthroughs should improve both the quality and the length of human life. Engineering of the environment will be unprecedented in its degree of intervention and control. Other technology trend effects may be less obvious to the public but in hindsight may be quite revolutionary. Fundamental changes in what and how we manufacture will produce unprecedented customization and fundamentally new products and capabilities.

Despite the inherent uncertainty in looking at future trends, a range of technological possibilities and effects are foreseeable and will depend on various enablers and barriers (see Table 3.1).

These revolutionary effects are not proceeding without issue. Various ethical, economic, legal, environmental, safety, and other social concerns and decisions must be addressed as the world's population comes to grip with the potential effect of these trends on their cultures and their lives. The most significant issues may be privacy, economic disparity, cultural threats (and reactions), and bioethics. In particular, issues such as eugenics, human cloning, and genetic modification invoke the strongest ethical and moral reactions. Understanding these issues is quite complex, since they both drive technology directions and influence each other in secondary and higher-order ways. Citizens and decisionmakers need to inform themselves about technology, assembling and analyzing these complex interactions to truly understand the

---

[3]See, for example, the discussion of regional concerns about culture and technology in Hundley et al. (2000 [212]).

debates surrounding technology. Such steps will prevent naive decisions, maximize technology's benefit given personal values, and identify inflection points at which decisions can have the desired effect without being negated by an unanalyzed issue.

Technology's promise is here today and will march forward. It will have widespread effects across the globe. Yet, the effects of the technology revolution will not be uniform, playing out differently on the global stage depending on acceptance, investment, and a variety of other decisions. There will be no turning back, however, since some societies will avail themselves of the revolution, and globalization will thus change the environment in which each society lives. The world is in for significant change as these advances play out on the global stage.

SUGGESTIONS FOR FURTHER READING

General Technology Trends

- "Research and Development in the New Millennium: Visions of Future Technologies." Special issue of *R&D Magazine*, Vol. 41, No. 7, June 1999.
- *Global Mega-Trends*, New Zealand Ministry of Research, Science & Technology, http://www.morst.govt.nz/foresight/info.folders/global/intro.html.
- "Visions of the 21st Century." *TIME*, http://www.time.com/time/reports/v21/home.html.

Biotechnology

- *Biotechnology: The Science and the Impact* (Conference Proceedings), Netherlands Congress Centre, the Hague, http://www.usemb.nl/bioproc.htm, January 20–21, 2000.
- "Global issues: biotechnology," U.S. Department of State, International Information Programs, http://usinfo.state.gov/topical/global/biotech/.
- *Introductory Guide to Biotechnology*. The Biotechnology Industry Organization (BIO) http://www.bio.org/aboutbio/guidetoc.html.
- "Biotechnology," Union of Concerned Scientists, http://www.ucsusa.org/agriculture/0biotechnology.html.
- Dennis, Carina, Richard Gallagher, and Philip Campbell (eds.), "The human genome," special issue on the human genome, *Nature*, Vol. 409, No. 6822, February 15, 2001.
- Jasny, Barbara R., and Donald Kennedy (eds.), "The human genome," special issue on the human genome, *Science*, Vol. 291, No. 5507, February 16, 2001.

Materials Technology

- Olson, Gregory B., "Designing a new material world," *Science*, Vol. 288, No. 5468, May 12, 2000, pp. 993–998.
- Good, Mary, "Designer materials," *R&D Magazine*, Vol. 41, No. 7, June 1999, pp. 76–77.
- Gupta, T. N., "Materials for the human habitat," *MRS Bulletin*, Vol. 25, No. 4, April 2000, pp. 60–63.
- *Smart Structures and Materials: Industrial and Commercial Applications of Smart Structures Technologies. Proceedings of SPIE,* Volumes 3044 (1997), 3326 (1998), and 3674 (1999). The International Society for Optical Engineering, Bellingham, Washington.
- The Intelligent Manufacturing Systems Initiative being pursued by Australia, Canada, The European Union, Japan, Switzerland, and the United States (with Korea about to be admitted) maintains a web page at http://www.ims.org.
- Kazmaier, P., and N. Chopra, "Bridging size scales with self-assembling supramolecular materials," *MRS Bulletin*, Vol. 25, No. 4, April 2000, pp. 30–35.
- Newnham, Robert E., and Ahmed Amin, "Smart Systems: Microphones, Fish Farming, and Beyond," *Chemtech*, Vol. 29, No. 12, December 1999, pp. 38–46.
- "Manufacturing a la carte: agile assembly lines, faster development cycles," *IEEE Spectrum*, special issue, Vol. 30, No. 9, September 1993.

Nanotechnology

- Coontz, Robert, and Phil Szuromi (eds.), "Issues in nanotechnology," *Science*, Vol. 290, No. 5496, special issue on nanotechnology, November 24, 2000, pp. 1523–1558.
- *National Nanotechnology Initiative: Leading to the Next Industrial Revolution,* Executive Office of the President of the United States, http://www.nano.gov/.
- *Nanostructure Science and Technology: A Worldwide Study,* National Science and Technology Council (NSTC), Committee on Technology and the Interagency Working Group on NanoScience, Engineering and Technology (IWGN), http://www.nano.gov/.
- Smalley, R. E., "Nanotechnology and the next 50 years," presentation to the University of Dallas Board of Councilors, http://cnst.rice.edu/, December 7, 1995.
- Freitas, Robert A., Jr., "Nanomedicine," *Nanomedicine FAQ*, www.foresight.org, January 2000.

FORESIGHT METHODS

1. Coates, J. F., "Foresight in Federal government policy making," *Futures Research Quarterly*, Vol. 1, 1985, pp. 29–53.

2. Martin, Ben R., and John Irvine, *Research Foresight: Priority-Setting in Science*, Pinter, London, 1989.

3. Larson, Eric V., "From forecast to foresight: lessons learned from a recent U.S. technology foresight activity," Keynote session, *Foresight at Crossroads Conference*, November 29–30, 1999.

GENERAL S&T FUTURE VISIONS

4. *Global Mega-Trends*, New Zealand Ministry of Research, Science & Technology, http://www.morst.govt.nz/foresight/info.folders/global/intro.html.

5. "Visions of the 21st Century," *TIME*, http://www.time.com/time/reports/v21/home.html.

6. Hammonds, Keith H., "The optimists have it right," *Business Week*, August 13, 1998.

7. Gross, Niel, and Otis Port, "The next wave for technology," *Business Week*, August 13, 1998.

8. Campbell, Philip, "Tales of the expected," *Nature*, Vol. 402 Supp., December 1999, pp. C7–C9.

9. "Simulating chemistry," R&D Research and Development in the New Millennium, *R&D Magazine*, Vol. 41, No. 7, June 1999, pp. 44–48.

10. Greenspan, A., "Maintaining economic vitality," Millennium Lecture Series, sponsored by the Gerald R. Ford Foundation and Grand Valley State University, Grand Rapids, Michigan, http://www.federalreserve.gov/boarddocs/speeches/1999/19990908.htm, September 8, 1999.

BIOTECHNOLOGY

11. "Biotechnology for the 21st century: New Horizons," Biotechnology Research Subcommittee, Committee on Fundamental Science, National Science and Technology Council, http://www.nal.usda.gov/bic/bio21/, July 1995.

12. "The biotech century," *Business Week*, March 10, 1997, pp. 79–92.

13. Lederberg, Joshua, "Science and technology: biology and biotechnology," *Social Research*, Vol. 64, No. 3, Fall 1997, pp. 1157–1161.

14. Zucker, Lynne G., Michael R. Darby, and Marilynn B. Brewer, "Intellectual human capital and the birth of U.S. biotechnology enterprises," *The American Economic Review*, Vol. 88, No. 1, March 1998, pp. 290–306.

15. "We are now starting the century of biology: already, genetic engineering is transforming medicine and agriculture—and that's just scratching the surface," *Business Week*, http://www.businessweek.com/1998/35/b3593020.htm, August 24–31, 1998.

16. Rotman, David, "The next biotech harvest," *MIT Technology Review*, September/October 1998.

17. Long, Clarisa, "Picture biotechnology: promises and problems," *The American Enterprise*, http://www.theamericanenterprise.org/taeso98p.htm, September 1, 1998, pp. 55–58.

18. Mironesco, Christine, "Parliamentary technology assessment of biotechnologies: a review of major TA reports in the European Union and the USA," *Science and Public Policy*, Vol. 25, No. 5, October 1998, pp. 327–342.

19. PricewaterhouseCoopers LLP, "Pharma 2005—an industrial revolution in R&D," 1998.

20. Gorman, Siobhan, "Future pharmers of America," *National Journal*, Vol. 31, No. 6, February 2, 1999, pp. 355–356.

21. Morton, Oliver, "First fruits of the new tree of knowledge," *Newsweek.com*, February 3, 1999.

22. Carey, John, Naomi Freundlich, Julia Flynn, and Neil Gross, "The biotech century—there's a revolution brewing in the lab, and the payoff will be breathtaking," *Business Week*, No. 3517, March 10, 1999, pp. 78–90.

23. Poste, George, "The conversion of genetics and computing: implications for medicine, society, and individual identity," Presentation to the Science and Technology Policy Institute, Summary by Danilo Pelletiere, www.rand.org/centers/stpi/newsci/Poste.html, April 19, 1999.

24. Pfeiffer, Eric W. (ed.), "Will biotech top the net?" *Forbes ASAP*, special issue on biotechnology, http://www.forbes.com/asap/99/0531/, May 31, 1999.

25. Logan, Toni, Evantheia Schibsted, Alex Frankel, Sally McGrane, and Suzie Amer, "Bioworlds: emerging pharma, it's all about drugs," *Forbes ASAP*, http://www.forbes.com/asap/99/0531/044.htm, May 31, 1999, pp. 44–57.

26. Caplan, Arthur, "Silence = disaster: to succeed biotech will have to answer many vexing ethical questions," *Forbes ASAP*, http://www.forbes.com/asap/99/0531/082.htm, May 31, 1999, pp. 82–84.

27. "Biotech mania," R&D Research and Development in the New Millennium, *R&D Magazine*, Vol. 41, No. 7, June 1999, pp. 22–27.

28. Gunter, Barrie, Julian Kinderlerer, and Deryck Beyleveld, "The media and public understanding of biotechnology: a survey of scientists and journalists," *Science Communication*, Vol. 20, No. 4, June 1999, pp. 373–394.

29. Stone, Amey, "This fund makes biotech bets a bit less risky," *Business Week*, July 12, 1999.

30. Thiel, Karl A., "Big picture biology," www.BioSpace.com/articles/, July 14, 1999.

31. Zorpette, Glenn, and Carol Ezzell (eds.), "Your bionic future," *Scientific American Presents*, September 1999.

32. "The third generation of pharmaceutical R&D introduction," Glaxo Wellcome, 21 October 1999.

33. Slavkin, Harold C., "Insights on human health: announcing the biotechnology century," National Institute of Dental & Craniofacial Research, www.nidr.nih.gov/slavkin/slav0999.htm, November 11, 1999.

34. "Industry in 2010: beyond the millennium," FT.com Life Sciences: Pharmaceuticals, http://www.ft.com/ftsurveys/q4b1a.htm, November 17, 1999.

35. "Biotech trends 100," *MIT Technology Review*, Vol. 102, No. 6, November/December 1999, pp. 91–92.

36. "Biotech on the move," *MIT Technology Review*, Vol. 102, No. 6, November/December 1999, pp. 67–69.

37. PricewaterhouseCoopers LLP, "Pharma 2005—silicon rally: the race to e-R&D," 1999.

38. "Biotech 2030: eight visions of the future," www.biospace.com/articles/, January 6, 2000.

39. "Systems biology in the post-genomics era," *Signals Magazine*, http://recap/coom.signalsmag.nsf/DP91D8DF, February 2, 2000.

40. Hapgood, Fred, "Garage biotech is here or just around the corner: will genetic modification for fun and profit become a homegrown industry?" *Civilization*, April/May 2000, pp. 46–51.

41. Biotechnology Industry Organization (BIO), *Introductory Guide to Biotechnology*, 2000, http://www.bio.org/aboutbio/guidetoc.html.

BIOSENSORS AND RELATED SENSORS

42. Schultz, Jerome S., "Biosensors," *Scientific American*, August 1991, pp. 64–69.

43. Scheller, F. W., F. Schubert, and J. Fedrowitz, *Frontiers in Biosensorics I: Fundamental Aspects*, and *II: Practical Applications*, Birkhäuser, Basel, 1997.

44. Dickinson, Todd A., Joel White, John S. Kauer, and David R. Walt, "Current trends in 'artificial-nose' technology," *Trends in Biotechnology*, Vol. 16, June 1998, pp. 250–258.

45. Simpson, Michael L., Gary S. Sayler, Bruce M. Applegate, Steven Ripp, David E. Nivens, Michael J. Paulus, and Gerald E. Jellison, Jr., "Bioluminescent-bioreporter integrated circuits form novel whole-cell biosensors," *Trends in Biotechnology*, Vol. 16, August 1998, pp. 332–338.

46. Giuliano, Kenneth A., and D. Lansing Taylor, "Fluorescent-protein biosensors: new tools for drug discovery," *Trends in Biotechnology*, Vol. 16, March 1998.

47. Hellinga, Homme W., and Jonathan S. Marvin, "Protein engineering and the development of generic biosensors," *Trends in Biotechnology*, Vol. 16, April 1998, pp. 183–189.

48. Marose, Stefan, Carsten Lindemann, Roland Ulber, and Thomas Scheper, "Optical sensor systems for bioprocess monitoring," *Trends in Biotechnology*, Vol. 17, January 1999, pp. 30–34.

GENOMICS

49. Coates, Joseph, F., John B. Mahaffie, and Andy Hines, "The Promise of Genetics," *The Futurist*, Vol. 31, No. 5, September–October 1997, pp. 18–22.

50. Strohman, Richard C., "Five stages of the human genome project," *Nature Biotechnology*, Vol. 17, February 1999, p. 112.

51. Naomi Freundlich, "Finding a cure in DNA?" *Business Week*, No. 3517, March 19, 1999, pp. 90–92.

52. "Human genome promise," *R&D Magazine*, Vol. 41, No. 7, June 1999, pp. 40–42.

53. Jasny, Barbara R., and Pamela J. Hines, "Genome prospecting," *Science*, Vol. 286, No. 5439, October 15, 1999, pp. 443–491.

54. Plomin, Robert, "Genetics and general cognitive ability," *Nature*, Vol. 402, Supp., December 16, 1999, pp. C25–C29.

55. Eisen, Jonathan, "Microbial and plant genomics," *Biotechnology: The Science and the Impact* (Conference Proceedings), Netherlands Congress Centre, the Hague, http://www.usemb.nl/bioproc.htm, January 20–21 2000.

56. Carrington, Damian, "How the code was cracked," *BBC News Online*, http://news.bbc.co.uk/hi/english/in_depth/sci_tech/2000/human_genome/newsid_760000/760849.stm, May 30, 2000.

57. Pennisi, Elizabeth, "Finally, the book of life and instructions for navigating it," *Science*, Vol. 288, No. 5475, June 30, 2000, pp. 2304–2307.

58. Dennis, Carina, Richard Gallagher, and Philip Campbell (eds.), "The human genome," special issue on the human genome, *Nature*, Vol. 409, No. 6822, February 15, 2001.

59. Baltimore, David, "Our genome unveiled," *Nature*, Vol. 409, No. 6822, February 15, 2001, pp. 814–816.

60. Aach, John, Martha L. Bulyk, George M. Church, Jason Comander, Adnan Derti, and Jay Shendure, "Computational comparison of two draft sequences of the human genome," *Nature*, Vol. 409, No. 6822, February 15, 2001, pp. 856–859.

61. International Human Genome Sequencing Consortium (IHGSC), "Initial sequencing and analysis of the human genome," *Nature*, Vol. 409, No. 6822, February 15, 2001, pp. 860–921.

62. Jasny, Barbara R., and Donald Kennedy (eds.), "The human genome," special issue on the human genome, *Science*, Vol. 291, No. 5507, February 16, 2001.

63. Galas, David J., "Making sense of the sequence," *Science*, Vol. 291, No. 5507, February 16, 2001, pp. 1257–1260.

64. Venter, J. Craig, et al., "The sequence of the human genome," *Science*, Vol. 291, No. 5507, February 16, 2001, pp. 1304–1351.

GENETICALLY MODIFIED FOODS

65. Evenson, Robert E., "Global and local implications of biotechnology and climate change for future food supplies," *Proceedings of the National Academies of Science USA*, Vol. 96, May 1999, pp. 5921–5928.

66. Butler, Declan, Tony Reichhardt, Alison Abbott, David Dickson, and Asako Saegusa, "Long-term effect of GM crops serves up food for thought," *Nature*, Vol. 398, No. 6729, April 22, 1999.

67. Vogt, Donna U., and Mickey Parish, "Food biotechnology in the United States: science, regulation, and issues," Congressional Research Service Report to Congress, http://usinfo.state.gov/topical/global/biotech/crsfood.htm, June 2, 1999.

68. Persley, G. J., and M. M. Lantin (eds.), *Agricultural Biotechnology and the Poor*, An International Conference on Biotechnology, Consultative Group on International Agricultural Research, http://www.cgiar.org/biotech/rep0100/contents.htm, October 21–22, 1999.

69. Haslberger, Alexander G., "Monitoring and labeling for genetically modified products," *Science*, Vol 287, No. 5452, January 21, 2000, pp. 431–432.

70. Somerville, Chris, "The genetic engineering of plants," *Biotechnology: the Science and the Impact* (Conference Proceedings), Netherlands Congress Centre, the Hague, http://www.usemb.nl/bioproc.htm, January 20–21, 2000.

71. Benbrook, Charles, "Who controls and who will benefit from plant genomics?" *The 2000 Genome Seminar: Genomic Revolution in the Fields: Facing the Needs of the New Millennium,* AAAS Annual Meeting, Washington D.C., February 19, 2000.

72. Langridge, William H.R., "Edible Vaccines," *Scientific American*, Vol. 283, No. 3, September 2000, pp. 66–71.

CLONING

73. Eiseman, Elisa, *Cloning Human Beings: Recent Scientific and Policy Developments,* RAND, MR-1099.0-NBAC, http://www.rand.org/publications/MR/MR1099.pdf, Santa Monica, California, August 1999.

74. Gurdon, J. B., and Alan Colman, "The future of cloning," *Nature*, Vol. 402, No. 6763, December 16, 1999, pp. 743–746.

75. Pennisi, E., and G. Vogel, "Animal cloning: clones: a hard act to follow," *Science*, Vol. 288, No. 5472, June 9, 2000, p. 1722–1727.

76. McLaren, Anne, "Cloning: pathways to a pluripotent future," *Science*, Vol. 288, No. 5472, June 9, 2000, p. 1775–1780.

77. Matzke, M. A., and A. J. M. Matzke, "Cloning problems don't surprise plant biologists," *Science*, Vol. 288, No. 5475, June 30, 2000, p. 2318.

78. Weiss, Rick, "Human cloning's 'numbers game,'" *Washington Post,* October 10, 2000, p. A01.

NEW ORGANISM CREATION AND MINIMAL GENOMES

79. Cho, Mildred K., David Magnus, Arthur L. Chaplain, and Daniel McGee, "Ethical considerations in synthesizing a minimal genome," *Science*, Vol. 286, No. 5447, December 10, 1999, pp. 2087–2090.

80. Hutchinson III, Clyde A., Scott N. Peterson, Steven R. Gill, Robin T. Cline, Owen White, Claire M. Frazer, Hamilton O. Smith, and J. Craig Venter, "Global

transponson mutagenesis and a minimal mycoplasma genome," *Science*, Vol. 286, No. 5447, December 10, 1999, pp. 2165–2169.

FLUORESCENT PROTEINS

81. Tsien, Roger Y., "Rosy dawn for fluorescent proteins," *Nature Biotechnology*, Vol.17, October 1999, pp. 956–957.

LAB TECHNIQUES/INSTRUMENTS

82. Müller-Gärtner, Hans-W., "Imaging techniques in the analysis of brain function and behaviour," *Trends in Biotechnology*, Vol. 16, No. 3., March 1998.

83. Thomas, Charles F., and John G. White, "Four-dimensional imaging: the exploration of space and time," *Trends in Biotechnology*, Vol. 16, April 1998, pp. 175–182.

84. Foster, Barbara, "MicroConvergence," *R&D Magazine*, Vol. 41, No. 8, July 1999, pp. 48–50.

85. Stoffel, James, "New imaging pathways," *R&D Magazine*, Vol. 41, No. 7, June 1999, pp. 80–82.

86. "Shrinking the 'universal sensor'," *R&D Magazine*, Vol. 41, No. 9, August 1999, pp. S53–S54.

87. Lindsay, Stuart, "AFM emerges as essential R&D tool," *R&D Magazine*, Vol. 41, No. 10, September 1999.

88. Costello, Catherine E., "Bioanalytic applications of mass spectrometry," *Current Opinion in Biotechnology*, Vol. 10, 1999, pp. 22–28.

COMBINATORIAL CHEMISTRY

89. Karet, Gail, "Combinatorial methods successful in solid-state catalyst discovery," *R&D Magazine*, Vol. 41, No. 11, October 1998, p. 67.

90. Pople, John, quoted in "Simulating chemistry," *R&D Magazine*, Vol. 41, No. 7, June 1999.

LAB-ON-A-CHIP

91. Schena, Mark, Renu A. Heller, Thomas P. Theriault, Ken Konrad, Eric Lachenmeier, and Ronald W. Davis, "Microarrays: biotechnology's discovery platform for functional genomics," *Trends in Biotechnology*, Vol. 16, July 1998, pp. 301–306.

92. Studt, Tim, "Development of microfluidic UHTS systems speeding up," *R&D Magazine*, Vol. 41, No. 2, February 1999, p. 43.

93. Hicks, Jennifer, "Genetics and drug discovery dominate microarray research," *R&D Magazine*, Vol. 41, No. 2, February 1999, pp. 28–33.

94. Marsili, Ray, "Lab-on-a-chip poised to revolutionize sample prep," *R&D Magazine*, Vol. 41, No. 2, February 1999, pp. 34–40.

95. "Fundamental changes ahead for lab instrumentation," *R&D Magazine*, February 1999, Vol. 41, No. 2, pp. 18–27.

96. Regnier, Fred E., Bing He, Shen Lin, and John Busse, "Chromatography and electrophoresis on chips: critical elements of future integrated, microfluidic analytical systems for life science." *Trends in Biotechnology*, Vol. 17, March 1999, pp. 101–106.

97. "Robotics speed drug discovery," *R&D Magazine*, Vol. 41, No. 9, August 1999, p. S57.

98. "Biochips perform genetic analyses rapidly and economically," *R&D Magazine*, Vol. 41, No. 9, August 1999, p. S54.

BIO-NANOTECHNOLOGY

99. Drexler, K. Eric, "Building molecular machine systems," *Trends in Biotechnology*, Vol. 17, January 1999, pp. 5–7.

100. Pum, Dietmar, and Uwe B. Sleytr, "The application of bacterial S-layers in molecular nanotechnology," *Trends in Biotechnology*, Vol. 17, January 1999, pp. 8–12.

101. Parkinson, John, and Richard Gordon, "Beyond micromachining: the potential of diatoms," *Trends in Biotechnology*, Vol. 17, May 1999, pp. 190–196.

102. Lee, Stephen C., "Biotechnology for nanotechnology," meeting report, *Trends in Biotechnology*, Vol. 16, June 1999, pp. 239–240.

103. Merkle, Ralph C., "Biotechnology as a route to nanotechnology," *Trends in Biotechnology*, Vol. 17, July 1999, pp. 271–274.

104. Kröger, Nils, Rainer Deutzmann, and Manfred Sumper, "Polycationic peptides from diatom biosilica that direct silica nanosphere formation," *Science*, Vol. 286, November 1999, pp. 1129–1131.

105. Amato, Ivan, "Reverse engineering the ceramic art of algae," *Science*, Vol. 286, November 1999, pp. 1059–1061.

106. Mirkin, C. A., "A DNA-based methodology for preparing nanocluster circuits, arrays, and diagnostic materials," *MRS Bulletin*, Vol. 25, No. 1, January 2000, pp. 43–54.

BIOMEDICAL ENGINEERING

107. Grundfest, Warren, "The future of biomedical engineering at UCLA," UCLA colloquium, March 16, 2000.

108. Chang, Thomas Ming Swi, "Artificially boosting the blood supply," *Chemistry & Industry*, April 17, 2000, pp. 281–285.

BIOLOGICAL SIMULATIONS

109. Noble, Denis, "Reduction and integration in understanding the heart," *The Limits of Reductionism in Biology*, Gregory Bock and Jamie Goode (eds.), Novartis Foundation Symposium, Vol. 213, John Wiley & Sons Ltd., 1998, pp. 56–68; discussion pp. 68–75.

110. Robbins-Roth, Cynthia, "The virtual body," *Forbes*, Vol. 162, No. 3, August 10, 1998, p. 109.

111. Buchanan, Mark, "The heart that just won't die," *New Scientist*, Vol. 161, No. 2178, March 20, 1999, pp. 24–28.

112. Normile, Dennis, "Building working cells '*in silico*,'" *Science*, Vol. 284, No. 5411, April 2, 1999, pp. 80–81.

113. Schaff J., and L. M. Loew, "The virtual cell," *Pacific Symposium on Biocomputing*, Vol. 4, 1999, pp. 228–239.

BIOINFORMATICS

114. *Trends Guide to Bioinformatics*, Elsevier Science, 1998.

115. Lim, Hwa A., and Tauseef R. Butt, "Bioinformatics takes charge," meeting report, *Trends in Biotechnology*, Vol. 16, March 1998.

116. Venkatesh, T. V., Benjamin Bowen, and Hwa A. Lim, "Bioformatics, pharma and farmers," meeting report, *Trends in Biotechnology*, Vol. 17, March 1999, pp. 85–88.

STEM CELLS

117. Shamblott, M. J., J. Axelman, S. Wang, E. M. Bugg, J. W. Littlefield, P. J. Donovan, P. D. Blumenthal, G. R. Huggins, and J. D. Gearhart, "Derivation of pluripotent stem cells from cultured human primordial germ cells," *Proceedings of the National Academy of Sciences USA*, Vol. 95, 1998, pp. 13726–13731.

118. Thomson, J. A., J. Itskovitz-Eldor, S. S. Shapiro, M. A. Waknitz, J. J. Swiergiel, V. S. Marshall, and J. M. Jones, "Embryonic stem cell lines derived from human blastocysts," *Science*, Vol. 282, No. 5391, November 6, 1998, pp. 1145–1147.

119. Couzin, Jennifer, "The promise and peril of stem cell research—scientists confront thorny ethical issues," *U.S. News & World Report*, May 31, 1999.

120. U.S. National Bioethics Advisory Commission, "Ethical issues in human stem cell research," http://bioethics.gov/pubs.html, September 1999.

121. McLaren, Anne, "Stem cells: golden opportunities with ethical baggage," *Science*, Vol. 288, No. 5472, June 9, 2000, p. 1778.

122. Allen, Arthur, "God and science," *Washington Post Magazine*, October 15, 2000, pp. 8–13, 27–32.

MATERIALS SCIENCE AND ENGINEERING

123. National Research Council, *Materials Science and Engineering in the 90s: Maintaining Competitiveness in the Age of Materials*, National Academy Press, Washington, D.C., 1989.

124. Good, Mary, "Designer materials," *R&D Magazine*, Vol. 41, No. 7, June 1999, pp. 76–77.

125. Arunachalam, V. S., "Materials challenges for the next century," *MRS Bulletin*, Vol. 25, No. 1, January 2000, pp. 55–56.

126. ASM International, "Millennium materials," Editorial, *Advanced Materials & Processes*, March 2000.

127. Gupta, T. N., "Materials for the human habitat," *MRS Bulletin*, Vol. 25, No. 4, April 2000, pp. 60–63.

128. Olson, Gregory B., "Designing a new material world," *Science*, Vol. 288, No. 5468, May 12, 2000, pp. 993–998.

BIOMATERIALS

129. Nadis, Steve, "We can rebuild you," *MIT Technology Review*, Vol. 100, No. 7, October 1997, pp. 16–18.

130. Bonassar, L. J., and C. A. Vacanti, "Tissue engineering: the first decade and beyond," *J. Cell. Biochem. Suppl.*, Vol. 30/31, 1998, pp. 297–303.

131. Aksay, I. A., and S. Weiner, "Biomaterials, is this really a field of research?" *Current Opinion in Solid State & Materials Science*, Vol. 3, 1998, pp. 219–220.

132. Colbert, Daniel T., and Richard E. Smalley, "Fullerene nanotubes for molecular electronics," *Trends in Biotechnology*, Vol. 17, February 1999, pp. 46–50.

133. McFarland, Eric W., and W. Henry Weinberg, "Combinatorial approaches to materials discovery," *Trends in Biotechnology*, Vol. 17, March 1999, pp. 107–115.

134. Garnett, M. C., et al., "Applications of novel biomaterials in colloidal drug delivery systems," *MRS Bulletin*, Vol. 24, No. 5, May 1999, pp. 49–56.

135. Reiss, J. G., and M. P. Krafft, "Fluorocarbons and fluorosurfactants for in vivo oxygen transport (blood substitutes), imaging, and drug delivery," *MRS Bulletin*, Vol. 24 No. 5, May 1999, pp. 42–48.

136. Glaev, Igor Y., and Bo Mattiasson, "'Smart' polymers and what they could do in biotechnology and medicine," *Trends in Biotechnology*, Vol. 17, No. 8, August 1999, pp. 335–340.

137. Ackerman, Robert K., "Futuristic materials inspired by biological counterparts," *Signal*, March 2000, pp. 37–41.

138. Temenoff, J. S., and A. G. Mikos, "Review: tissue engineering for regeneration of articular cartilage," *Biomaterials*, Vol. 21, 2000, pp. 431–440.

139. Hench, L. L., "Medical materials for the next millennium," *MRS Bulletin*, Vol. 24, No. 5, May 1999, pp. 13–19 (see also http://www.anl.gov/OPA/news96/news961203.html).

140. For additional information on tissue engineering, see: http://www.pittsburgh-tissue.net, http://www.whitaker.org, http://www.advancedtissue.com, http://www.organogenesis.com, http://www.integra-ls.com, and http://www.isotis.com.

RAPID PROTOTYPING AND ROBOTICS

141. "How hot lasers are taming titanium," *Fortune*, Industrial Management and Technology Edition, 21 February 2000, quoted at http://www.aerometcorp.com/aeromentnews.htm.

142. Rapid prototyping web page and links were found at http://www.cc.utah.edu/~asn8200/rapid.html#COM.

SMART MATERIALS AND STRUCTURES

143. Haertling, G. H., "RAINBOW ceramics—a new type of ultra-high-displacement actuator," *American Ceramic Society Bulletin*, Vol. 73, 1994, pp. 93–96.

144. Humbeeck, J. Van, D. Reynaerts, and J. Peirs, "New opportunities for shape memory alloys for actuators, biomedical engineering, and smart materials," *Materials Technology*, Vol. 11, No. 2, 1996, pp. 55–61.

145. Sater, Janet M. (ed.), *Smart Structures and Materials 1997: Industrial and Commercial Applications of Smart Structures Technologies*, Proceedings of SPIE, Volume 3044, The International Society for Optical Engineering, Bellingham, Washington, May 1997.

146. Newnham, R. E., "Molecular mechanisms in smart materials," *MRS Bulletin*, Vol. 22, No. 5, May 1997, pp. 20–34.

147. Shahinpoor, M., Y. Bar-Cohen, J. O. Simpson and J. Smith, "Ionic polymer-metal composites (IPMCs) as biomimetic sensors, actuators and artificial muscles—a review," *Smart Mater. Struct.*, Vol. 7, 1998, pp. R15–R30.

148. Sater, Janet M. (ed.), *Smart Structures and Materials 1998: Industrial and Commercial Applications of Smart Structures Technologies*. Proceedings of SPIE, Volume 3326, The International Society for Optical Engineering, Bellingham, Washington, June 1998.

149. Jacobs, Jack H. (ed.), *Smart Structures and Materials 1999: Industrial and Commercial Applications of Smart Structures Technologies*, Proceedings of SPIE, Volume 3674, The International Society for Optical Engineering, Bellingham, Washington, July 1999.

150. Newnham, Robert E., and Ahmed Amin, "Smart systems: Microphones, fish farming, and beyond," *Chemtech*, Vol. 29, No. 12, December 1999, pp. 38–46.

151. Bar-Cohen, Y., "Electroactive polymers as artificial muscles—capabilities, potentials and challenges," Keynote Presentation at the *Robotics 2000* and *Space 2000* International Conferences (International Conference and Exposition on Engineering, Construction, Operations, and Business in Space, collocated with the International Conference and Exposition on Robotics for Challenging Situations and Environments), Albuquerque, New Mexico, February 28–March 2, 2000, http://www.spaceandrobotics.org. The site includes references to other electroactive polymer and related robotics websites.

152. Wool, Richard P., "Polymer science: A material fix," *Nature*, Vol. 409, No. 6822, February 15, 2001, pp. 773–774.

153. White, S. R., N. R. Sottos, P. H. Geubelle, J. S. Moore, M. R. Kessler, S. R. Sriram, E. N. Brown, and S. Viswanathan, "Autonomic healing of polymer composites," *Nature*, Vol. 409, No. 6822, February 15, 2001, pp. 794–797.

NANOMATERIALS

154. Alivisatos, P., "Electrical studies of semiconductor nanocrystal colloids," *MRS Bulletin*, Vol. 23 No. 2, February 1998, pp. 19–23.

155. Smalley, R. E., "Nanotech growth," *R&D Magazine*, Vol. 41, No. 7, June 1999, pp. 34–37.

156. Chen, T., N. N. Thadhami, and J. M. Hampikian, "The effects of nanostructure on the strengthening of NiAl," *High-Temperature Ordered Intermetallic Alloys Symposium VIII*, Materials Research Society Symposium Proceedings, Vol. 552, Materials Research Society, Pittsburgh, Pennsylvania, 1999.

157. Koch, C. C., D. G. Morris, K. Lu, and A. Inoue, "Ductility of nanostructured materials," *MRS Bulletin*, Vol. 24, No. 2, 1999, pp. 54–58.

NANOTUBES

158. Bockrath, M., et al., "Single electron transport in ropes of carbon nanotubes," *Science*, Vol. 275, 1997, p. 1922.

159. Zhu, W., C. Bower, O. Zhou, G. Kochanski, and S. Jin, "Large current density from carbon nanotube field emitters," *Applied Physics Letters*, Vol. 75, No. 6, 1999, pp. 873–875.

160. Lerner, Eric J., "Putting nanotubes to work," *The Industrial Physicist*, American Institute of Physics, December 1999, p. 22–25.

MOLECULAR MANUFACTURING

161. Service, Robert F., "New reaction promises nanotubes by the Kilo," *Science*, Vol. 290, No. 5490, October 13, 2000, pp. 246–247.

162. Drexler, K. Eric, *Engines of Creation: The Coming Era of Nanotechnology*, Anchor Books/Doubleday, New York, 1987.

163. Drexler, K. Eric, *Nanosystems: Molecular Machinery, Manufacturing, and Computation*, John Wiley and Sons, New York, 1992.

164. Nelson, Max, and Calvin Shipbaugh, *The Potential of Nanotechnology for Molecular Manufacturing*, RAND, MR-615-RC, Santa Monica, California, 1995.

165. Crandall, B. C. (ed.), *Nanotechnology: Molecular Speculations on Global Abundance*, The MIT Press, Cambridge, Massachusetts, 1996.

166. Timp, Gregory (ed.), *Nanotechnology*, Springer Verlag, New York, 1999.

167. Voss, David, "Moses of the nanoworld: Eric Drexler led the way," *MIT Technology Review*, Vol. 102, No. 2, March/April, 1999.

168. Zachary, G. Pascal, "Nano-hype," *MIT Technology Review*, Vol. 103, No. 1, January/February 2000, p. 39.

NANOTECHNOLOGY

169. Smalley, R. E., "Nanotechnology and the next 50 years," Presentation to the University of Dallas Board of Councilors, http://cnst.rice.edu/, December 7, 1995.

170. Smith, Charles G., "Computation without current," *Science*, Vol. 284, No. 5412, April 1999, p. 274.

171. Siegel, Richard W., Evelyn Hu, Mihail C. Roco, *Nanostructure Science and Technology (A World-Wide Study): R&D Status and Trends in Nanoparticles,*

Nanostructured Materials and Nanodevices, National Science and Technology Council (NSTC) Committee on Technology and The Interagency Working Group on NanoScience, Engineering and Technology (IWGN), Dordrecht: Kluwer Academic, 1999 (also available at http://itri.loyola.edu/nano/IWGN.Worldwide.Study/).

172. "Nanotech growth," Research and development in the new millennium, *R&D Magazine*, Vol. 41, No. 7, June 1999.

173. McWhorter, Paul J., "The role of nanotechnology in the second silicon revolution," Testimony before the U.S. House of Representatives Committee on Science, June 22, 1999.

174. Merkle, Ralph, C., "Nanotechnology: the coming revolution in manufacturing," Testimony before the U.S. House of Representatives Committee on Science, June 22, 1999.

175. Wong, Eugene, "Nanoscale science and technology: opportunities for the twenty-first century," Testimony before the U.S. House of Representatives Committee on Science, June 22, 1999.

176. Smalley, R. E., "Nanotechnology," Testimony before the U.S. House of Representatives Committee on Science, June 22, 1999.

177. Freitas, Robert A. Jr., "Nanomedicine," *Nanomedicine FAQ*, www.foresight.org, January 2000.

178. "National Nanotechnology Initiative: Leading to the Next Industrial Revolution," White House press release, http://www.whitehouse.gov/WH/New/html/20000121_4.html, January 21, 2000.

179. *National Nanotechnology Initiative: Leading to the Next Industrial Revolution*, Executive Office of the President of the United States, http://www.nano.gov, February 7, 2000.

180. Rennie, John, "Nanotech reality," *Science*, Vol. 282, No. 6, June 2000, p. 8.

181. Coontz, Robert, and Phil Szuromi (eds.), "Issues in nanotechnology," special issue on nanotechnology, *Science*, Vol. 290, No. 5496, November 24, 2000, pp. 1523–1558.

MOLECULAR ELECTRONICS

182. P. S. Weiss, "Are single molecular wires conducting?" *Science*, Vol. 271, 1996, pp. 1705–1707.

183. Cuberes, M. T., et al., "Room temperature repositioning of individual C60 molecules at Cu steps: Operation of a molecular counting device," *Appl. Phys. Lett.*, Vol. 69, 1996, p. 3016.

184. Credi, A., V. Balzani, S. J. Langford, and J. F. Stoddart, "Logic operations at the molecular level. An XOR gate based on a molecular machine," *J. Am. Chem. Soc.*, Vol. 119, 1997, p. 2679.

185. Collins, P. G., A. Zettl, H. Bando, A. Thess, and R. E. Smalley, "Nanotube nanodevice," *Science*, Vol. 278, p. 100.

186. Heath, James R., Philip J. Kuekes, Gregory S. Snider, and R. Stanley Williams, "A defect-tolerant computer architecture: opportunities for nanotechnology," *Science*, Vol. 280, No. 5370, June 1998, pp. 1716–1721.

187. Collier, C. P., E. W. Wong, M. Belohradsky, F. M. Raymo, J. F. Stoddart, P. J. Kuekes, R. S. Williams, and J. R. Heath, "Electronically configurable molecular-based logic gates," *Science*, Vol. 285, No. 5426, 1999, pp. 391–394.

188. Chen, J., M. A. Reed, A. M. Rawlett, and J. M. Tour, "Large on-off ratios and negative differential resistance in a molecular electronic device," *Science*, Vol. 286, 1999, pp. 1550–1552.

NANOFABRICATED CHIPS

189. Packen, Paul, "Pushing the limits," *Science*, Vol. 285, No. 5436, 24 September, 1999, pp. 2079–81.

190. SEMATECH, *International Technology Roadmap for Semiconductors*, 1999.

QUANTUM COMPUTING

191. Shor, P., "Algorithms for quantum computation: Discrete logarithms and factoring," *Proc. 35th Ann. Symp. Foundations of Computer Science*, Vol. 124, 1994.

192. Bennet, C. H., "Quantum information and computing," *Physics Today*, Vol. 48, No. 10, October 1995, pp. 24–30.

193. DiVincenzo, D., "Quantum computation," *Science*, Vol. 270, 1995, p. 255.

194. Gershenfeld, N., and I. L. Chuang, "Bulk spin resonance quantum computation," *Science*, Vol. 275, 1997, p. 350.

195. Sohn, Lydia L., "A quantum leap for electronics," *Nature*, Vol. 394, No. 6689, July 1998.

196. Birnbaum, J., and R. S. Williams, "Physics and the information revolution," *Physics Today*, Vol. 53, No. 1, January 2000, pp. 38–42.

BIO-COMPUTING

197. Adleman, L., "Molecular computation of solutions to combinatorial problems," *Science*, Vol. 266, 1994, p. 1021.

198. Alivisatos, A. P., et al., "Organization of 'nanocrystal molecules' using DNA, " *Nature*, Vol. 382, 1996, p. 609.

199. "Computing with DNA," *Scientific American*, Vol. 279, 1998, p. 34.

200. Tomita, M., K. Hashimoto, K. Takahashi, Y. Matsuzaki, R. Matsushima, K. Saito, K. Yugi, F. Miyoshi, H. Nakano, S. Tanida, and T. S. Shimizu, "E-CELL project overview: towards integrative simulation of cellular processes," *Genome Informatics Workshop 1998*, Tokyo, Japan, 10–12 December 1998, http://www.genome.ad.jp/manuscripts/GIW98/Poster/GIW98P02.pdf.

201. Tomita, M., K. Hashimoto, K. Takahashi, T. S. Shimizu, Y. Matsuzaki, F. Miyoshi, K. Saito, S. Tanida, K. Yugi, J. C. Venter, and C. A. Hutchison, 3rd, "E-CELL: software environment for whole-cell simulation," *Bioinformatics*, Vol. 15, No. 1, January 15, 1999, http://www.sfc.keio.ac.jp/~mt/mt-lab/publications/abs/tomita99.html, pp. 72–84.

MEMS

202. Marshall, Sid, "New applications emerging as MEMS technology advances," *R&D Magazine*, Vol. 41, No. 8, July 1998, pp. 32–37.

203. Picraux, S. Tom, and Paul J. McWhorter, "The broad sweep of integrated microsystems," *IEEE Spectrum*, December 1998, pp. 24–33.

204. Sasaki, Satoshi, and Isao Karube, "The development of microfabricated biocatalytic fuel cells," *Trends in Biotechnology*, Vol. 17, February 1999, pp. 50–52.

205. *Micromachine Devices*, Vol. 4, No.6, June 1999.

206. Karet, Gail, "Integrated approach simplifies MEMS design," *R&D Magazine*, Vol. 41, No. 8, July 1999, p. 41.

207. Marshall, Sid, "Industry roadmap planned for microsystems technology," *R&D Magazine*, Vol. 41, No. 8, July 1999, pp. 44–45.

NANOSENSORS

208. Dong, L. F., et al., "Gas sensing properties of nano-ZnO prepared by arc plasma method," *Nanostruct. Mater.*, Vol. 8, 1997, p. 815.

209. Duncan, R., "Polymer therapeutics for tumor specific delivery," *Chemistry and Industry*, Vol. 7, 1997, pp.262–264.

210. Iddan, G., G. Meron, and P. Swain, "Medical engineering: Wireless capsule endoscopy," *Nature*, Vol. 405, No. 6785, May 25, 2000, p. 417.

INFORMATION TECHNOLOGY VISIONS

211. Smarr, Larry, "Digital fabric," *R&D Magazine*, Vol. 41, No. 7, June 1999, pp. 50–54.

212. Hundley, Richard O., Robert H. Anderson, Tora K. Bikson, James A. Dewar, Jerrold Green, Martin Libicki, and C. Richard Neu, *The Global Course of the Information Revolution: Political, Economic, and Social Consequences: Proceedings of an International Conference*, RAND, CF-154-NIC, http://www.rand.org/publications/CF/CF154/, Santa Monica, California, 2000.

213. Anderson, Robert H., Philip S. Antón, Steven C. Bankes, Tora K. Bikson, Jonathan P. Caulkins, Peter J. Denning, James A. Dewar, Richard O. Hundley, and C. Richard Neu, *The Global Course of the Information Revolution: Technology Trends: Proceedings of an International Conference*, RAND, CF-157-NIC, http://www.rand.org/publications/CF/CF157/, Santa Monica, California, 2000.

TECHNOLOGIES FOR SPACE

214. Luu, Kim, and Maurice Martin, "GSFC shuttle payload design workshop for the university nanosatellite program," Overview and NASA Safety Workshop, AFOSR and DARPA University Nanosatellite Program, http://www.nanosat.usu.edu/presentations/afpayload/index.html, July 27, 1999. See also http://www.nanosat.usu.edu/ for general information on the nanosatellite program.

215. Beardsley, Tim, "The way to go in space," *Scientific American*, February 1999, pp. 81–97.

216. Marshall, Sid, "MEMS growth reflected in space instrumentation," *R&D Magazine*, Vol. 41, No. 8, July 1999, pp. 37–40.

GLOBALIZATION

217. Friedman, Thomas L., *The Lexus and the Olive Tree*, Anchor Books, New York, April 2000.

LEGAL ISSUES

218. Walter, Carrie F., "Beyond the Harvard Mouse: current patent practice and the necessity of clear guidelines in biotechnology patent law," *Indiana Law Journal*, Vol. 73, No. 3, http://www.law.indiana.edu/ilj/v73/no3/walter.html, Summer 1998.